Copernicus Books

Sparking Curiosity and Explaining the World

Drawing inspiration from their Renaissance namesake, Copernicus books revolve around scientific curiosity and discovery. Authored by experts from around the world, our books strive to break down barriers and make scientific knowledge more accessible to the public, tackling modern concepts and technologies in a nontechnical and engaging way. Copernicus books are always written with the lay reader in mind, offering introductory forays into different fields to show how the world of science is transforming our daily lives. From astronomy to medicine, business to biology, you will find herein an enriching collection of literature that answers your questions and inspires you to ask even more.

John G. Cramer

How to Live Much Longer

The Mitochondrial DNA Connection

John G. Cramer
Seattle, WA, USA

ISSN 2731-8982 ISSN 2731-8990 (electronic)
Copernicus Books
ISBN 978-3-032-17740-7 ISBN 978-3-032-17741-4 (eBook)
https://doi.org/10.1007/978-3-032-17741-4

Disclaimer: Many longevity interventions described in this book, and particularly the high-volume mitochondrial transplantation procedure, have not been tested on human subjects. Caution is advised. Any such treatment is considered risky and possibly dangerous until it has been fully tested. The author, as a volunteer for the first mitochondrial transplantation tests, has fully accepted the responsibility for this decision and understands the significant possibility of negative outcomes and consequences.

This Springer imprint is published by the registered company Springer Nature Switzerland AG
The registered company address is: Gewerbestrasse 11, 6330 Cham, Switzerland

But at my back I always hear
Time's wingèd chariot hurrying near;
And yonder all before us lie
Deserts of vast eternity.

Andrew Marvell
(1621–1678)
"To His Coy Mistress"

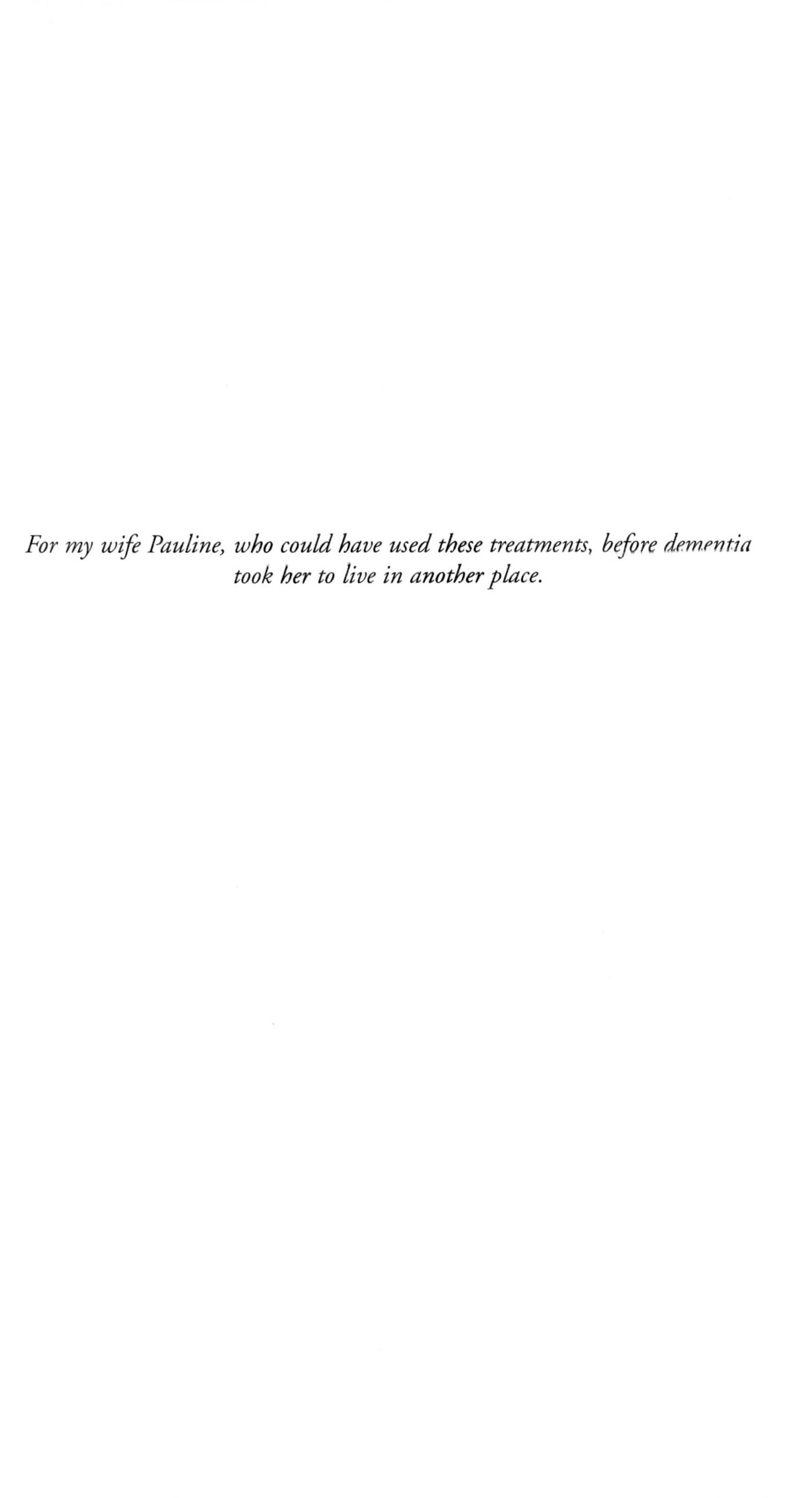

For my wife Pauline, who could have used these treatments, before dementia took her to live in another place.

Preface

About the Author

This book is about human aging and its root causes and underlying mechanisms. It is also about a *Great Adventure in Human Longevity*, in which I am presently playing a part, and that may, 30 years from now, allow me to become the oldest "young" human being on the planet.

I am a retired Professor of Physics at the University of Washington, who worked for six decades in the research fields of experimental and theoretical nuclear and particle physics, finally using the world's largest particle accelerators at labs like Brookhaven and CERN to make nuclear collision event fireballs that replicate the super-hot plasma in the first microseconds of the Big Bang.

And so, it is natural for the skeptical reader to ask why I should be venturing so far "out of my lane" to write this book on a subject that involves the remote fields of biochemistry, molecular biology, and medicine rather than fundamental physics. Let me explain.

(The author, John G. Cramer *Credit* JGC)

I was born on October 24, 1934, so my age at this writing is 91 years and some months, and growing. I retired from my tenured teaching faculty position in the Department of Physics of the University of Washington on January 1, 2010, at age 75, and I phased out my DOE-funded physics research activities over the following decade, at the same time writing the nonfiction popular-level physics book *The Quantum Handshake* (Springer, 2016) to present my transactional interpretation of quantum mechanics to a general audience and writing my third "hard science" SF novel, *Fermi's Question* (Baen, 2024), the sequel to my second SF novel, *Einstein's Bridge* (AvonNova, 1997). My most recent physics journal publication, "Symmetry, Transactions, and the Mechanism of Wave Function Collapse", was the last of my over 300 physics journal publications. It appeared in 2020 in the physics journal ***Symmetry 12*** (8), 1373 (2020); https://doi.org/10.3390/sym12081373.

With increasing age and more available time, the sharp intellectual focus of my physics instincts somehow turned to age-related questions: *Just what is human aging? What are its underlying driving mechanisms? What can be done to slow, stop, or reverse it? What are its limits?*

If you know where to look, the Internet science archives are full of research papers about aging (all with a rather high "understanding barrier" due to their extensive use of arcane bio-medical jargon). As I began to read these and to learn the unfamiliar bio-lingo, I came to realize that, despite the dozen or so theories of aging in the literature, the ideas are all over the map, with theories pointing in all directions. *There is presently no informed general consensus on the true root causes of human aging.*

As I will discuss later in this book, I feel that scientific researchers are either attempting to get at the fundamental mechanism behind their observations by using the scientific method to make and test hypotheses and models, or else they are simply doing plausible experiments with the tools at hand in particular situations, making observations and cataloging the results. We throw a switch and in another room a light bulb illuminates. Do we determine the structure of the switch, trace the wiring, analyze the glowing filament, find the power source, and model the circuit using mathematics, or do we simply record the action and result? My observation of the bio-literature is that, in the current state of aging research, most of the work falls in the latter category, and the authors do not, perhaps cannot, concern themselves with root causes.

That is quite understandable. Medical and biological experiments, particularly on human subjects, are done in situations in which it is very difficult to control all of the unrelated variables and under conditions that are almost impossible to simulate on a computer or to accurately describe with a mathematical theory or model. This leads to experimental results with error bars and variances so large as to make most physicists turn away in disgust (or cry). Getting at fundamental mechanisms is difficult, and the path to valid understanding is littered with mistaken ideas and wrong-headed missteps. The "not invented here" syndrome as a criterion for rejecting new ideas is also rampant.

Cataloged experimental results are useful and certainly should be published. In fact, unlike physics journals, some of the leading biomedical journals and archives will *only* accept papers for publication if they contain such data (no stand-alone insights, ideas, or theories allowed).

We do what we can, so I continued my practice of reading several aging papers a day, making a few notes, and visiting and having discussions with University of Washington "South Campus" faculty members engaged in biological research relevant to aging (see Acknowledgments). I entered into email discussions with like-minded individuals online, asking many questions (I'm good at that) of people and also of AIs, and forming (and sometimes later rejecting) new ideas, theories, and opinions. When something seemed to be working, e.g., an "intervention" that made recipients feel younger or that made old mice act friskier and live longer, I wanted to know not only what was done, but *why* it had worked, and the size of the effect it produced.

As a physicist, I know that energy is conserved in even the most complicated situations, so a physics-based approach to aging snapped into place. I have attempted to "follow the energy", i.e., to try to understand complex biological processes from the point of view of energy requirements, supply, exchange, and balance. This approach has led me along a crooked path to *mitochondria*, the many tiny bacteria-like organelles residing within each cell that provide it with all of its energy, internally supply spare parts and lubricants to its spinning energy supply machinery, and play a role in intercellular communication and control. Mitochondria are descended from ancient oxygen-breathing bacteria that took up symbiotic residence in one of our hydrogen-breathing single-cell ancestors about two billion years ago. Mitochondria were first discovered in 1857, were given their name in 1898, were associated with cellular energy production in the 1950s, and were discovered to have their own independent small ring of DNA, ribosomes, and unique genetic code in the 1960s. By 2013 "Mitochondrial dysfunction" was recognized to be enough of a health problem that it was listed as one of the nine Hallmarks of Aging (see Chap. 5 below).

However, only in the past few years has this particular aging hallmark been seriously considered as *the possible central driver behind all of human aging*. There remains considerable controversy and debate about whether something as complex as human aging could possibly have just one single root cause. However, as research on the effects of cumulative mitochondrial deletion mutation damage emerges (see Appendix B below), that possibility is looking more and more likely. I decided to write this book because my physics training has led me to focus on mitochondrial energy as an organizing concept that ties together all of aging's other Hallmarks, and because this easily-explained perspective deserves far more attention in the popular arena of human health and aging.

I believe that energy availability and energy balance in the cellular world are the keys to understanding the process of human aging. I hope that our present understanding and application of this fundamental energy concept are only the beginning of a new dawn, heralding a more complete understanding of how life on this planet actually works.

I should mention that as a practicing physicist I have a certain approach to new problems. The famous physicist Ernest Rutherford, discoverer of the atomic nucleus and winner of the 1908 Nobel Prize in Chemistry, is reputed to have said: "All science is either physics or stamp collecting." What I think he meant by that statement, which many non-physicists find rather offensive, is that some scientists are simply collecting specimens or trying things within their capabilities, just to see what happens, with no focus on root causes or mechanisms, and they are satisfied with cataloging their results. Examples: "*We collected and identified a previously unknown species of tree toad on a certain date in habitat X at location Y,*" or "*We gave the patient medication Z and his blood pressure dropped into the normal range.*" These activities are worthwhile and valuable "stamps", in that they can lead others to generalizations or to the treatment of problems, and they contribute to the large volume of human records that future scientists can use for their own purposes and theories.

However, such cataloging does not in itself lead to *understanding*, i.e., to discovering the underlying root causes of phenomena like amphibian evolution or high blood pressure. That's where the "physics" part comes in. The way that physics is supposed to be done (and sometimes is) is to construct a theory that depicts the causal relationships within a particular phenomenon, use that theory to predict new untested and unknown results, and then consult result databases, make new observations, or perform new experiments to check the validation or falsification of that prediction. In other words, to seek understanding rather than a catalog of results. Here, we will attempt to *understand* human aging.

One additional point: Something that I have noticed in my recent readings of a large number of journal publications touching on the biology of aging is that there is a certain amount of "magical thinking" going on there, what below is called the "Free Ride Fallacy". For example, *one infuses this protein or "knocks out" that gene and some interesting or desirable outcome follows.* What is missed is the causal chain that leads from the action to the result. In particular, a question not often asked in biology is how much *energy* is required to produce the observed effect, where did that energy come from, and would it be available in all circumstances? A common practice in physics is to *follow the energy*. It would be wise for practitioners in biology to adopt that practice more widely. It allows the use of logic and reasoning that can lead to new insights and understanding (preferred over cataloging).

Goals of This Book

This book about aging and its treatment is intended for the interested non-specialist who would like to understand the underlying biological causes of and possible remedies for human aging. As we will see, aging is a very complicated process, but we will try to keep it simple, providing clear explanations and examples while avoiding as much as possible of the arcane terminology favored by the molecular biology community, with its fondness for Latin terms and specialist lingo. As examples, we will often say that a gene *produces* (rather than *expresses*) a protein or that a particular cell is a *fat cell* (rather than an *adipocyte*). For the targeted level of understanding, it will often be necessary to go into some technical detail. However, I will do my best to keep things simple. Where biotechnical terms are needed, I will try to add parenthetical explanations to provide them with more immediate meaning.

I fully expect that, with a few exceptions, my readers under the age of about 60 will find the material in this book to be somewhat distant, theoretical, and perhaps beside the point, because their lives are presently going well enough and old age, its causes, and its treatments seem rather distant and remote, off on the far horizon. Readers in their 70s will find these things a bit more immediate and relevant, but probably still a bit on the theoretical side. For that reason, the readers in their 80s and beyond are my real targets. The ticking clock of age-related disability confronts them all, and they will not need to be convinced that aging is an immediate problem that they urgently need to understand and to deal with however and whenever possible.

This book describes, at minimum, my own hard-won understanding of what human aging actually is. It includes some discussion of possible and practical ways of treating and reversing aging and its effects. Unfortunately for most of us, the best and most effective of these ways are still 5–10 years off in the future. If we have only a few years of healthy life left, there is not much room for an extended waiting period. That is why, in Chaps. 6 and 7 below, I have described in specific detail a number of longevity interventions that are available *now* and that, if used effectively, can perhaps enable us to continue our normal lives for long enough for the stronger interventions like large-volume mitochondrial transplantation and epigenetic reprogramming to become generally available and within our economic reach.

It will be a desperate race against time, but a race that we have some possibility of winning.

Seattle, USA John G. Cramer

Acknowledgments The author wishes to thank Steve Bard, Gary Hudson, Matt Scholz, Johnny Adams, Tom Benson, David Gobel, Steve Horvath, Scott Kennedy, Matt Kaeberlein, David Marcinek, Bonita J. Brewer, and Jonathan Wanagat, in no particular order, for very helpful discussions, emails, comments, paper and data sharing, and general encouragement.

Steve Bard, starting in the 1980s, introduced me to the use of anti-aging supplements and to the ideas of life extension, telomeres, and telomerase. Gary Hudson and Matt Scholz, co-founders of Oisin Bio, introduced me to the concepts of senescent cells and the importance of clearing them as one ages, and they were very tolerant in dealing with several "hot" ideas that I presented to them. Tom Benson, CEO of Mitrix Bio, introduced me to the key role of mitochondria in human aging, became my co-author on a scientific paper and several white papers, recruited me to be among the first volunteers for the projected Mitrix trial of large-volume mitochondrial transplantation on humans, and has been of invaluable assistance in the preparation of this book. Johnny Adams, co-founder and moderator of the online Gerontology Research Group (GRG), provided me directly and indirectly with countless useful interactions and concepts about human aging. David Gobel, founder of the Methuselah Foundation, arranged for me to become a "Chronaut" and provided stimulating HBOT treatments. Steve Horvath, discoverer of several epigenetic clocks and tests, was very generous with his time in answering my many questions by email. Scott Kennedy, Matt Kaeberlein, David Marcinek, and Bonita J. Brewer, all faculty members and bio-researchers at the "South Campus" of the University of Washington, have helped me with many online interactions and in-person visits. Jonathan Wanagat (UCLA) has very generously provided the data from their work that I used to generate Fig. B.1.

John G. Cramer

Competing Interests The author has made modest investments in four bio-startups, Oisin Bio, OncoSenX, Turn Bio, and Mitrix Bio, which are relevant to the content of this book.

Executive Summary

This book proposes that the complex phenomenon of human aging is fundamentally the result of a cellular energy crisis. The central argument is that aging's primary driver is the progressive accumulation of damage to the DNA within mitochondria (mtDNA), the essential energy producers in our cells. This perspective reframes many of the recognized "Hallmarks of Aging," viewing them not as independent phenomena but as downstream consequences of this basic energy deficit. The book traces this energy vulnerability back to a critical moment in evolutionary history, termed a "Faustian Bargain," an instant when one of our single-celled ancestors acquired abundant energy through mitochondrial symbiosis. In exchange, it sacrificed its cellular immortality. The trade-off introduced a new vulnerability: the ticking time bomb of inevitable mtDNA failure, which leads to aging and death. This vulnerability was passed on to us.

While various current longevity interventions offer modest, short-term benefits, they are identified as merely addressing the symptoms of aging without resolving the underlying energy shortage that lies at their source. The author argues that achieving genuine age reversal requires more robust and fundamental measures. The primary proposed solution is the emerging technology of high-volume mitochondrial transplantation. This procedure aims to directly replenish the body's cells with healthy, functional mitochondria, thereby restoring youthful cellular energy production. This fundamental energy restoration is presented as a crucial first step, which could potentially reawaken dormant cellular repair mechanisms. It should also enable the safe and effective use of other advanced rejuvenation therapies like senolytics and epigenetic reprogramming—processes that are themselves highly energy-intensive.

A successful rollout of this lifespan-extending technology should bring with it profound biomedical transformations, starting with revolutionary treatments for mitochondrial genetic diseases and new protocols for emergency medical care, particularly for treating strokes and heart failure. However, it is also likely to introduce unprecedented societal, economic, and ethical challenges. These include issues of sports performance enhancement (mitochondrial "doping"), equitable access to rejuvenating treatments, potential social stratification, and disruptions to the labor market and the population age balance, as well as possible added strain on the environment. Consequently, navigating the coming age-reversal revolution will require thoughtful and intelligent modification of existing societal structures and ethical frameworks, leading humanity to redefine its relationship with mortality, progress, and life itself.

Contents

List of Figures

1

Going Gentle into That Good Night? An Analogy

1.1 Into That Good Night

The great Welsh poet and author Dylan Thomas [Th52] admonished us:

Do not go gentle into that good night,

Old age should burn and rave at close of day;

Rage, rage against the dying of the light.

Thomas' own departure into That Good Night came in 1953 at age 39, following bouts of heavy drinking and carousing in the face of an ongoing battle with his own failing lungs, and that response could be taken as his chosen version of the rage. My own version, seeking at age 91 to rage against the dying of the light, perhaps my own burning and raving at close of day, takes the form of the writing of this book. In reading it, you should take that for what it's worth.

My other, possibly more effective, way of raging against the dying of the light is to volunteer to be an early participant in the world's first human trials of mitochondrial transplantation, the preliminary stages of which are now being conducted by Mitrix Bio. As we will see later in this book, this not-yet-tested or FDA-approved human age reversal intervention promises to be the key to human longevity by providing new mitochondrial energy to the aging cells, in which the damaged mitochondria are failing at an exponentially

J. G. Cramer, *How to Live Much Longer*, Copernicus Books,
https://doi.org/10.1007/978-3-032-17741-4_1

growing rate. The cells of the aged are coping with a severe and growing energy shortage.

At age 91, I am the oldest volunteer scheduled to participate in this test of mitochondrial transplantation. As such, in another 30 years I will perhaps be confronted with the possibility of becoming the oldest living human being, who is continuing to exist and thrive well beyond the present 122 year limit of human aging. That's a very interesting and challenging prospect. I view it as embarking on a *great adventure in human longevity, in which there is the distinct possibility that I may become the oldest "young" person in the history of the human race.*

1.2 An Analogy Illustrating the Mitochondrial View of Aging

We will take a deep dive into the relationship between *mitochondria*, the cadre of busy little bacteria-like energy-producers that live and work within each human cell, and the phenomenon of human aging, the progressive degeneration of our mental and physical capabilities as we grow older, leading inevitably to our enfeeblement and death. We will begin this look at what's going on in human aging with a simple analogy (Fig. 1.1):

> Human genes are like a group of blue-collar industrial workers who produce assembly tools, parts, supplies, and major components for the Big Machine (that's us). Among the tens of thousands of these workers, about half of them are busily making needed tools, parts, and supplies. The other half stand idle, waiting patiently on the sidelines until their skills are needed. Their work assignments and schedules are complicated and change slowly under orders from an unseen supervising controller. Over time, these changes tend to side-line workers with high-energy-consuming jobs. Most of these workers (we'll call them the NUGies), live and work in elegant houses in a gated community, carefully protected from workplace and environmental damage by a high wall, a tough security force, and a top-of-the-line medical staff that can effectively treat most of the on-the-job damage and injuries that they sometimes receive.
>
> But on the other side of the tracks, deep in the polluted industrial zone, there are a hundred or so work groups of 37 workers each (we'll call them the MUGies), who are situated in a much more hazardous environment. Their jobs are to produce assembly tools, replacement parts, and lubricants for the supremely important fuel-burning energy-generating turbine engines themselves, the crucial infrastructural devices that power the entire Big Machine. Since the MUGies must work very closely with those engines, they are assigned

> to live and labor next to them, in trailer parks in the middle of the highly polluted industrial zone, where the energy production operation is actively spewing out damaging toxic byproducts. These workers have a minimal protective wall, less security, and a medical staff consisting of one minimally-trained guy who can provide 1st Aid.
>
> Not surprisingly, the MUGies have an active working life that is shorter than that of the NUGies (perhaps 20 times shorter). More important, although the MUGies are a smaller group, their energy-related jobs are so important that when their working productivity falls off or stops, an area-wide energy drought ensues, the NUGies lose their energy-expensive healthcare and accumulate more damage too, and the entire Big Machine sputters, slows down, declines, and eventually grinds to a stop.

Here, the NUGies (nucleus-unit genes) represent all the active and inactive genes of our nuclear DNA, many spooled onto protective histones and organized into chromosomes for more protection, surrounded by the protective wall of the cell nucleus, guarded by the aggressive immune system, and provided with a wide range of effective repair mechanisms. About half of these genes are actively involved in protein production and other vital cellular functions. The other half are "silenced" by CpG methylation (see Chap. 5), either because they have already made their contributions during our infancy, childhood, and youth, or because their contributions are not yet needed. As our age progresses, this pattern of active/inactive genes changes through a process called *epigenetic programming*. Its control underpinnings are not well understood, but they seem to be related to the availability of energy and other resources and to silencing non-essential genes, particularly those that place large demands on scarce resources, in favor of low-demand genes that perform similar functions.

The MUGies (mitochondrial-unit genes) are the 37 always-active genes encoded in our mitochondrial DNA (mtDNA), which in humans is a ring of DNA comprised of exactly 16,569 base pairs. There are 13 protein-coding genes and 24 RNA support genes. These genes are surrounded by only a minimal nucleoid wall, are open and fully accessible on the looping mtDNA ring, with only minimal protection from invading oxidizing molecules and toxins, and are provided with a replication enzyme that sporadically fails and stumbles over double DNA breaks, and a repair mechanism, inherited from primitive bacterial precursors, that addresses only a few types of single-point mutation damage and that frequently makes repair mistakes.

For the human genes that control our bodies, these are two different DNA worlds, one well protected and the other fairly hostile and dangerous. The

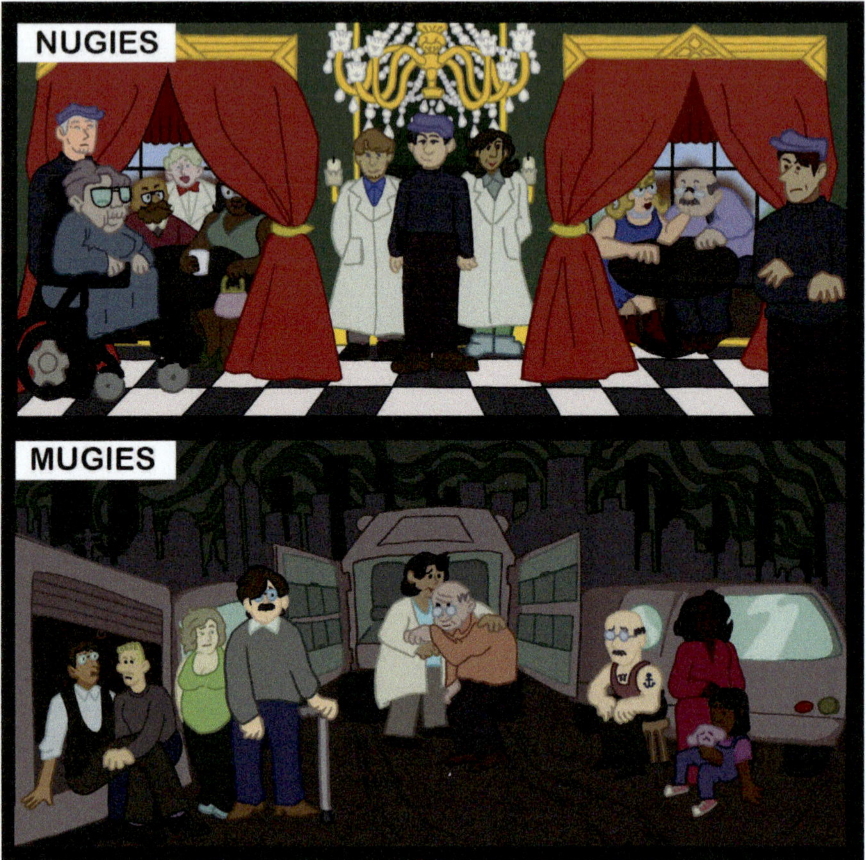

Fig. 1.1 The NUGies and the MUGies. *Credit* Ratio Hartwell

mtDNA accumulates damage around 20 times faster than the nuclear DNA, leading to an age-related exponential falloff in ATP energy production and to a growing shortage of the ATP energy that must be available to allow cells to perform their functions and to repair damage. This cellular energy depletion is the key weak link that is thought to be a principal driver of age-related disease and decline, perhaps the single root cause of all human aging.

Reference

[Th52] Dylan Thomas, ***Collected Poems, 1934–1952***, London: Dent (1952).

2

Ancient History: The Origin of Earth Life and Its Consequences

Instead of taking the current biological situation with cellular aging and human response as a given, it is useful to review our current understanding of just how we humans, and Earth life in general, came to be in our present situation. In particular, we would like to answer the questions of how and why our pretty blue planet has any life on it at all. In this chapter, we will consider the origins of life on Earth and the consequences of the way in which it came to be.

2.1 RNA Forms and Self-replication Begins

There was a time, long ago, when the emerging planet Earth, having condensed out of the debris from Big Bang leftovers and previous generations of exploding stars and supernovae, was in the violent turmoil of early formation, and there was *no life at all* on the developing planet. The Hadean Era, 4.6–4.0 billion years ago, was a violent time of tectonic plate shifting and rearrangement, extensive volcanic activity, and frequent bombardment from above by meteorites and asteroids [Le03]. There was even one colossal collision with an object that astronomers have named Theia, a Mars-size planetoid, that in a collision, supplied much extra mass to the developing Earth and scattered debris in orbit that, over time, coalesced to form the Moon (Fig. 2.1).

Astronomers call this period of many collisions from space *The Late Heavy Bombardment.* It was a relatively short geological period in which many massive planetesimals from the outer Solar System were somehow dislodged

J. G. Cramer, *How to Live Much Longer*, Copernicus Books, https://doi.org/10.1007/978-3-032-17741-4_2

Fig. 2.1 Collision of planetoid Theia with Earth, creating the Moon. *Credit* NASA/JPL-Caltech/T. Pyle, public domain

from stable orbits, perhaps by the large gravitational perturbation from a passing star or by a brief Jupiter-Saturn orbital resonance. The result of this disturbance was that some fraction of the outer planetesimals were deflected inward to bombard Earth and the other developing planets of the inner Solar System. The presence of the large amount of water on today's Earth is perhaps a consequence of heavy bombardment during this period by icy asteroids dislodged from orbits well outside the Solar System's "ice line", beyond which the objects are cold enough for water ice to condense and collect.

Is Earth a unique place? Among all of our galaxy's extra-solar Earth-like planets orbiting other Sun-like stars, such a violent late bombardment event during planetary formation must have been quite rare. This supports the "Rare Earth" hypothesis [Wa00] that our Earth, with its abundant water, atmospheric oxygen, large season-stabilizing Moon, and abundant life, may be a very rare planet in our galaxy and a very special place indeed. We should take care of it!

Earth's atmosphere during the Hadean Era was not breathable. It was mainly nitrogen, with some carbon dioxide and smaller amounts of free hydrogen, methane, and water vapor, but it contained *no free oxygen at all*. In some regions, Hadean Earth became cool enough that the atmospheric water vapor could rain down to form pools and lakes, as today's great oceans began to emerge. This chaotic and unstable environment somehow fostered the first progress toward the emergence of life, starting with the appearance of the first RNA molecules in the chemical-rich waters of Earth's developing oceans.

2.1.1 The RNA World

RNA (ribonucleic acid) is a very special organic molecule, described in much more detail in Appendix A below. It has a complicated "string of beads" structure (see Fig. A.3 below) that can contain information. It came into existence on Earth only after the planet had cooled enough for oceans to form, for things to settle down, and for the warm soup that was the ocean's salty water, with its load of dissolved chemicals, to began combining the atoms and simple molecules it contained and to brew up new more complex molecules. The emergence of RNA and the development of organizing interactions between RNA molecules are thought to be the key intermediate steps in the remarkable transition from lifeless inert matter to Earth's first self-replicating chemical systems.

The "RNA World" hypothesis, first proposed in 1962 by MIT Prof. Alexander Rich (1924–2015), suggests that RNA molecules formed rather early as the Earth was cooling and that somehow some of these RNA molecules developed the capability to *make new copies of themselves* [Ri62, Gi86, Ne13]. This RNA self-replication development occurred well before the appearance of DNA or any of the complex peptides and proteins that formed later. Self-replicating RNA provided an essential bridge that led to the ultimate development of true life. It led to the development of more complex self-replicating chemical-based systems.

2.1.2 Help from the Asteroids

The time when primordial RNA first appeared on Earth coincides suggestively with the last stages of the Late Heavy Bombardment described above. In the context of RNA formation, a type of jet black carbon-containing asteroids (carbonaceous chondrites) that participated in the bombardment are particularly relevant. They have been found to contain sizable quantities of amino acids and most of the nucleotides needed to form RNA [Pa21]. Curiously, the carbon-containing chondrites from space can supply only three of the four nucleobases that are needed to form RNA. Cytosine, a key RNA building block, is missing.

So, if the critical RNA component cytosine was *not* supplied from space, how could full RNA molecules have formed on Earth? In 1959, Urey and

Miller showed that the chemical reactions that follow a lightning strike on Earth's developing oceans could trigger the formation of complicated organic molecules, including nucleobases like cytosine [Mi59]. Also, the active hydrothermal vents deep beneath the ocean's surface spew out mineral-rich fluids. These vents, with their steep thermal and chemical gradients, could also have provided many needed organic molecules. Probably the first RNA was formed from component nucleotides provided by *both* asteroid bombardment and the lightning-induced and other chemical processes brewing in the Earth's developing oceans [La04]. RNA formation was a key development on the long road that led to Earth life.

2.2 Protocells: Earth Life Begins

The RNA World era ended when more complex self-replicating cell-like structures began to form and to compete for resources. Chemical processes in the Hadean Era oceans and asteroids from space had supplied RNA component nucleobases and amino acids in abundance. Long RNA sequences can fold themselves into complex shapes, and some of these intricately shaped molecules can become effective chemical machines and catalysts. One of the new RNA-built chemical machines was a primitive form of what we now call a *ribosome.* It can read a string of RNA, like a punched paper tape reader, and use the read-in sequence of triple nucleotide groups as a guide for combining amino acids, one by one, to assemble the complex molecular structures that we call *peptides* and *proteins.* Peptides are the shortest proteins, a string of less than about 50 amino acids, and they normally act as signaling molecules like hormones and neurotransmitters. The longer proteins can fold themselves into even more complex molecular machines, forming enzymes, transporters, and structural components. About 3.8 billion years ago, the development of such complex proteins led to the formation of the first *protocells* (Fig. 2.2).

Protocells were the starting point for the evolution of true cellular life [Mo24]. We have learned over the years that all of the life forms on Earth possess several common characteristics: (1) the use of a universal DNA-based

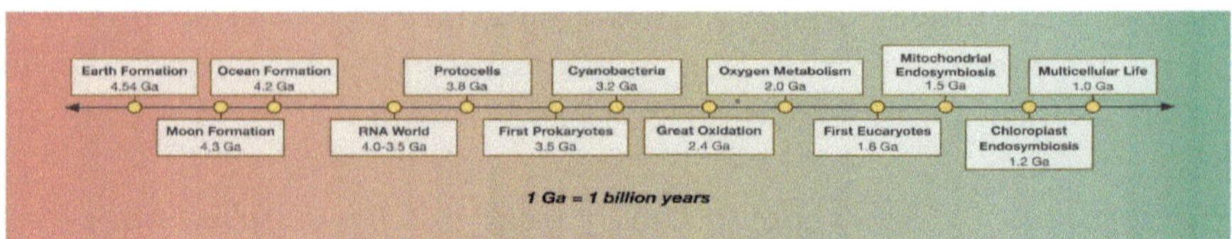

Fig. 2.2 Timeline for the development of life on earth. *Credit* Kathryn Cramer

"GCAT" three-letter genetic code (see Appendix A); (2) the use of a nearly universal messenger-RNA-based "GCAU" protein code, read by ribosome machinery to assemble proteins from amino acids; (3) the use of exclusively the *same* 20 amino acids from among the more than 500 naturally occurring amino acids to assemble all of the proteins; (4) the use of only left-handed amino acids to assemble the proteins, although their naturally-occurring mirror-image right-handed amino acid equivalents are also available; and (5) the use almost universally of the ATP molecule (adenosine triphosphate, or $C_{10}H_{16}N_5O_{13}P_3$) as the common energy currency of life. An ATP molecule cocks back one of its three phosphate molecules like a rock in a slingshot and, on command, shoots it out to supply the energy (about 0.33 electron-volts) used by all cellular life forms.

Earth's first protocells had all five of these characteristics, and they passed them on to their descendants (including us humans). Protocells had emerged from the intense RNA World activity in Earth's cooling oceans as more complex stepping stones in the evolution of life. Protocells did *not* contain fully functional DNA genetic material, but instead used simpler molecules like RNA to replicate. They had some rudimentary hydrogen-based metabolic processes, but they lacked the complexity and efficiency of a true cellular metabolism. They had primitive membrane walls made of simple lipids or other molecules that were less effective than the walls of true cells for through-the-wall substance exchange. The early protocells were out-competed by their better constructed descendants (discussed below), and so no protocells survive today for us to examine and analyze.

2.3 Prokaryotes: The First Cells

As the millennia passed in Earth's oceans, about 3.5 billion years ago the protocells evolved into the first true cells, the *prokaryotes* (here we will call them *P cells* to avoid arcane bio-terminology). P cells were simple small single-cell organisms, as shown in Fig. 2.3. They resembled the bacteria of today, which are their direct descendants. They replicated by cell division. The first P cells had the protection of a surrounding cell wall. Inside that wall were circular rings of DNA and some supporting ribosomes and transcription enzymes, but few other specialized organelles and *no cell nucleus at all*. The ring DNA of the first P cells is estimated to have had around 2.5 million base pairs containing around 2,600 genes. Present-day P cells have about twice as many base pairs and genes in their DNA and are often equipped with rotating whip-like flagella used for locomotion (which some of the first P cells may

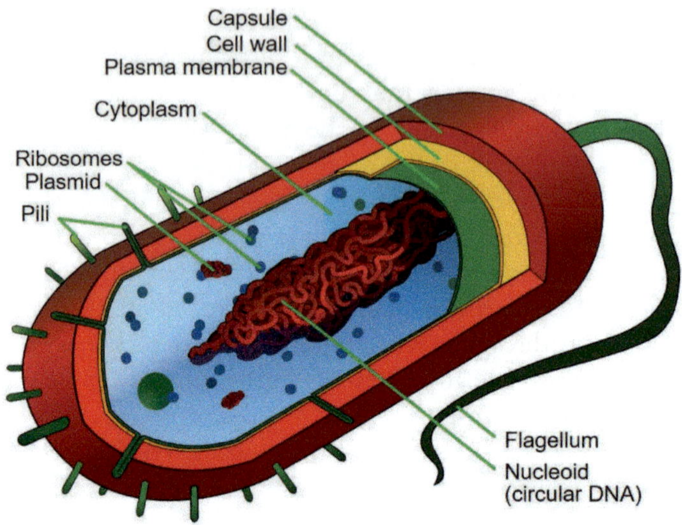

Fig. 2.3 A prokaryote cell. *Credit* LadyofHats, Public domain, via Wikimedia Commons

also have had). These P cells evolved into all of the present-day bacteria and all of the methane-producing archaea.

2.4 The Arrival of Photosynthesis and the Great Oxidation

Some of the ancient P cells lived on the surface of the oceans, where all of the abundant electromagnetic energy of sunlight was essentially going to waste. Then, about 3.2 billion years ago, one of them (which we identify as the first *cyanobacterium*) evolved a way to tap into sunlight's abundant energy using the process of *photosynthesis*, channeling the energy of the light to forcibly combine abundant carbon dioxide and water into glucose and free oxygen.

This bacterial innovator thrived, filling Earth's oceans with its progeny and irreversibly changing Earth's atmosphere. The long-term result of the new widespread photosynthesis was that Earth's atmosphere took on an increasing component of oxygen, coupled with shrinking components of free hydrogen gas, methane, and carbon dioxide.

That era's standard P cell metabolism of old had operated by taking in hydrogen and glucose and converting these to *acetate* ($C_2H_3O_2^-$) and ATP. Inputs of hydrogen gas, water, carbon dioxide, and carbon monoxide from the atmosphere or sub-ocean combined to liberate free acetate molecules and to supply ATP for cell energy. But this metabolic process depended on a supply of free hydrogen gas in the atmosphere or dissolved in the oceans, and that was rapidly disappearing.

About 2.4 billion years ago, all of that free oxygen being excreted by the thriving cyanobacteria began to thoroughly pollute Earth's atmosphere and oceans, killing some of the more vulnerable hydrogen-breathing life forms and combining with and reducing the available free hydrogen gas, in a geological event that is called the *Great Oxidation.*

For a time, all of that newly liberated free oxygen was going to waste as a potential cellular energy resource. Then, about 2 billion years ago, some enterprising P cell bacterium evolved enough to modify its metabolism so that it could *use* this new oxygen energy resource. Its former hydrogen-in-acetate-out metabolism was transformed to an oxygen-breathing metabolism that produced ATP and carbon dioxide from inputs of oxygen and glucose. *Oxygen breathers* had arrived on the Earth. These new little oxygen-breathing P cells were very successful in Earth's changing oceans and multiplied almost without limit.

2.5 Bigger and Better Cells: Eukaryotes Arrive

Evolution of life in Earth's oceans marched on, and about 1.8 billion years ago the first *eukaryotes,* far more advanced single-cell organisms, emerged. (Here we will call them *E cells*.) As illustrated in Fig. 2.4, E cells had a more complex internal structure and were much larger than their P cell ancestors. They had a substantial outer cell wall containing organelles for the entrance and exit of molecules, and an internal membrane-walled cell nucleus that held all of its very large single copy of its DNA. The DNA, along with translation, replication, and repair enzymes, was walled off within the cell nucleus, and much of the DNA was protected from random damage by being spooled onto histones and bundled into chromosomes. It was also protected by an aggressive immune system. The first E cells also had many other specialized organelles, formed from combined protein and RNA components, that performed many needed cell functions.

However, these first E cells *did not have mitochondria.* Further, although the oxygen content of the atmosphere was rising, it is probable that the first E cells were *not* oxygen breathers, but instead still used the old hydrogen-in-acetate-out metabolism of their earlier P cell ancestors. These E cells later evolved into all modern plants, fungi, and animals (including us humans).

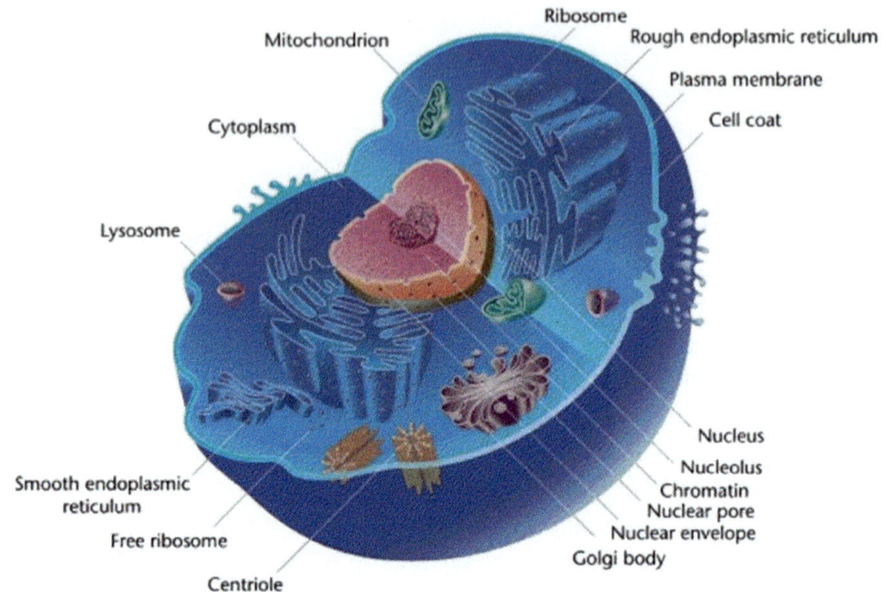

Fig. 2.4 A eukaryote cell. *Credit* Mediran, Public domain, via Wikimedia Commons

For a time, the giant well-organized E cells were metabolically left behind. They were vastly superior in their overall cellular design, but they were saddled with an obsolescent hydrogen-breathing metabolism, at a time when the global atmospheric hydrogen supply was shrinking, and they were coping with a very limited supply of ATP energy as they tried to compete for resources in Earth's oceans with their small and vigorous oxygen-breathing P cell rivals.

2.6 "The Capture Miracle": An E Cell Becomes a Supercell

Then, about 1.5 billion years ago, a game-changing E cell event occurred that we will call "*The Capture Miracle*". Some random little oxygen-breathing P cell wandered too close to a big hydrogen-breathing E cell and was engulfed and captured into the E cell's cytoplasm. But instead of being broken down there into useful protein parts and consumed as food, as would normally have happened, the captive P cell somehow reproduced itself and established a mutually-beneficial symbiotic relationship with its E cell captor. This was their deal: the E cell landlord would provide its new tenants with a safe home, glucose and other nutrients, and abundant oxygen. In exchange, the P cell occupants would go all out in producing large quantities of ATP for cell-wide energy use.

The lucky E cell, with the aid of its new P cell helpers, thus became what might be called a *supercell*, for it now was oxygen-breathing, had a superior design, and had available far more ATP energy for competition and expansion in Earth's oceans than its unmodified E and P cell rivals. It easily won all of the competitions and took over as the ocean's dominant single-cell life form. In the following Cambrian Era, about a billion years ago, that newly boosted E-cell energy powered the amazingly rapid development of the myriad of new and varied multi-cellular creatures that emerged to compete in and populate the Earth's oceans. That original energy-enhanced E cell was the ancestor of all plant, fungal, and animal life existing on the Earth today, including humanity.

The E cell's captured P-cell helpers have evolved into today's mitochondria, which in their current form are no longer able to live freely on their own, or even to survive for long outside the cytoplasm of their E cell hosts. The modern mitochondrion now retains only a small part of its original DNA, with the rest having moved to the better-protected cell nucleus. In mtDNA there remain only 13 genes coding for proteins, 2 genes for making its own ribosomes, and 22 genes for making transfer RNA that brings any of the standard 20 amino acids to the local ribosome for protein assembly.

However, the three-letter coding sequence used by today's mtDNA for adding a particular amino acid to the growing protein is *slightly different* from the code used by the cell's nuclear DNA. For example, the sequence TGA in human nuclear DNA is a "stop code" that terminates the assembly of a protein. In human mtDNA, the same sequence instead codes for adding the amino acid *tryptophan.* The origin of these coding differences is not well understood, but it is probably the result of some mtDNA mutation that occurred around a billion years ago. We note that modern bacteria and archaea use the *same* three-letter code as human nuclear DNA. For mysterious reasons, human mitochondrial DNA uses a slightly different three-letter coding sequence from the nearly universal one that is standard everywhere else in the bio-world.

2.7 Chloroplasts: A Second Capture

The plant-life equivalent of the E cell's "Capture Miracle" event occurred once again about 1.2 billion years ago for the plant-type E cells that now contained their own mitochondria. One of these plant E cells captured a passing photosynthetic cyanobacterium and incorporated it into its metabolism, a symbiosis providing the host cell with the ability to perform internal photosynthesis, i.e., to convert sunlight into chemical energy (glucose) while releasing oxygen as a byproduct. Over time, the captured cyanobacteria evolved into the plant cell's chlorophyll-containing *chloroplasts*, the foundation of all modern plant photosynthesis. All of today's plants evolved from that first photosynthetically enhanced plant E cell.

2.8 Slimming Down the mtDNA: Evolution in Action

Originally, the local DNA rings of the newly captured mitochondria contained all of the protein and RNA coding instructions needed for the production of new mitochondria, just as had their ancestral P-cell bacterial counterparts. However, over the passing eons that followed the capture event, the captured P cell's mitochondrial DNA genome was "slimmed down" from around 2.5 million base pairs, coding for around 2,600 genes to around 16.6 thousand base pairs coding for only 37 genes, with only 13 of these genes actually coding for the proteins that are components of the mitochondrial ATP production bio-machinery. Under evolutionary pressure, most of the important genetic coding needed for implementing mitochondrial structure and components had moved from the local mtDNA to the better-protected nuclear DNA of the host E cell.

There is a good reason for this evolution-driven gene migration: the mtDNA is located in a rather hostile and toxic environment, employs a somewhat error prone replication system, and has poor damage and error repair support. It is located in close proximity to the important and highly-active ATP production machinery, the current versions of which can crank out several hundred ATP molecules per second from each healthy Complex V site (see Fig. 4.2) within the mitochondrion. As a byproduct of that activity, particularly after minor damage occurs, the production process generates some reactive oxidizing species (ROS), oxidizing molecules that can damage nearby mtDNA and support enzymes and interfere with the mtDNA replication process that operates continuously within the mitochondrial matrix.

This hostile mitochondrial environment is responsible for the unrepaired point-substitution and sequence-deletion mutations that accumulate over time in mtDNA. This damage is estimated to grow **about 20 times faster** than similar damage in the cell's nuclear DNA. The high mtDNA mutation rate creates evolutionary pressure to move all but the most difficult-to-move and immediately needed DNA genetic coding to the more protected nuclear DNA environment. As noted above, in humans only 13 protein-coding genes and 24 RNA support genes have been left behind in the mtDNA. In contrast, around 1,100 mitochondria-related genes, all of which presumably had once resided in the first mtDNA, have now relocated to the cell's nuclear DNA.

This raises an interesting question: *Why did this process of gene relocation stop with 13 protein-coding genes left behind in the mtDNA?* Those 13 mtDNA genes were probably slow to relocate because the proteins they produce are all quite *hydrophobic*, meaning that they have a strong aversion to water. This would make it difficult to move through the cell's water-based cytoplasm surrounding the cell's ribosome to the mitochondrial interior (where they are components of the ATP-production machinery). That trip is not easy for hydrophobic proteins, but perhaps not impossible. A few mitochondrial genes coding for other hydrophobic proteins have already relocated to the nuclear DNA. However, these hydrophobic proteins are components of the mitochondrial wall membranes, meaning that they would not need to travel very far in the cytoplasm or to penetrate deep into the mitochondrial interior for installation.

Aubrey de Grey has offered a possible explanation for why the 13 vulnerable genes resisted evolutionary pressure and remain even now in the mtDNA, while all of the other mitochondrial genes have moved. As discussed above, all contemporary bacteria use the *same* nearly universal three-letter protein assembly code as that used by human nuclear DNA, while our mitochondria (which are cousins of those same bacteria) use a *different* code. It seems likely that around a billion years ago, some random inheritable mutation must have permanently changed the mtDNA's protein assembly code from that used by the nuclear DNA. With the establishment of these two incompatible protein assembly codes, the migration of protein-coding genes from mtDNA to nuclear DNA would have become much more difficult. The gene transfer process must have halted then, leaving behind the 13 slow-to-move genes coding for hydrophobic proteins, which today are still stuck in the mtDNA and subject to a much higher mutation and damage rate.

In the evolution of mitochondria-bearing E cells, there have been parallel but independent developments in the DNA-repair mechanisms for mitochondrial and for nuclear DNA, with those for nuclear DNA being far more effective. Perhaps this happened because of greater evolutionary pressure and because highly evolved E cells were the starting point. Mitochondria, descended from more primitive bacteria-like P cells, do provide base excision repair to fix oxidative mtDNA lesions, capable of restoring a base that ROS has removed from the chain, but they do not provide repair for most of the other ways in which mtDNA can be damaged.

For example, one common and devastating form of genome damage, particularly for aging cells, is the *deletion.* During the replication of a long DNA sequence, some random event, perhaps triggered by a radiation-induced gene double break or an oxidation event, can cause the replication enzyme to "skip" a length of the DNA sequence that can be thousands of base pairs long, only to resume replication somewhere well downstream, rather like an old vinyl disc record player skipping a groove. This creates a *deletion mutation*, which will be missing a long base pair sequence that is likely to eliminate or damage several essential mtDNA genes. For nuclear DNA, there are repair mechanisms for fixing this type of damage, but the only ways the mitochondrion has of dealing with such deletion damage are removal through mitophagy (see below) or ejecting an entire nucleoid capsule containing the deletion-damaged mtDNA ring, along with perhaps several other mtDNA rings and their support enzymes.

As discussed in Appendix B below, the ways in which mtDNA damage builds are rather complicated and differ among mutation types. Point mutation damage to mtDNA builds *linearly*, while deletion damage builds *exponentially*, with a doubling time in young adults of around 11.8 years and in the very old of around 3.06 years. Both types of mutations are responsible for the much higher rate of damage to mtDNA than to nuclear DNA. The consequence of this mitochondrial vs. nuclear difference is that, over time, the mtDNA accumulates unrepaired damage much more rapidly than does nuclear DNA. This is partially compensated for by the fact that cells contain only a single copy of their nuclear DNA but contain many mitochondria, each containing 2–10 or more copies of their mtDNA rings. However, when the majority of the mtDNA rings have damage, this advantage is lost. More is better, but it is not good enough.

In summary, the damage increases much faster in mtDNA than in nuclear DNA, with the result that the worn-out critical components of the mitochondrial ATP production machinery are not being replaced with new components and lubricants that they need for efficient ATP production. Consequently, over time, the ATP production machinery slows down and ultimately comes to a halt, depriving the myriad of vital cellular processes of the ATP energy they require to function normally. This is thought to be the very essence of human aging.

2.9 Energy and Aging: The Faustian Bargain of the Supercell

From one interesting perspective, the ancient symbiotic transition from a normal E cell to a mitochondria-bearing supercell was a *Faustian Bargain.* Acquiring the new and abundant mitochondrial source of ATP energy for the E cell brought with it a *ticking time bomb*, the progressively accumulating damage to the mtDNA in the newly acquired mitochondria and the escalating age-dependent cell and organ-wide ATP energy shortage that the growing mtDNA damage would inevitably cause.

Before the Capture Miracle, the primordial E cells were essentially immortal, splitting and replicating their well-protected and well-maintained DNA so that they could freely and repeatedly undergo cell division, making two cells from one over and over again, as they flourished in Earth's oceans. But after the P cell capture event, their mitochondria-bearing E cell descen dants were "on the clock", functioning and dividing for only a limited time period, for only as long as their supporting mtDNA remained sufficiently undamaged to sustain a viable supply of energy for their continued functioning and survival needs.

In other words, the price of having abundant energy in cellular youth was *to give up cell immortality*, and, in exchange, to suffer the inevitable growing shortage of energy in cellular old age, leading to decline and eventual cell death. Our primordial E cell ancestor bought into the prospect of an active and energetic youth, at the price of an inevitable slide into decrepitude, old age, and death. It has passed that bargain along to us. The human lifespan would perhaps be 20 times longer if progressive mtDNA damage were not bringing it to an early halt.

Gaining energy by losing immortality, or at least a longer lifespan, was a Faustian Bargain indeed (Fig. 2.5).

Fig. 2.5 The Faustian Bargain. *Credit* Ratio Hartwell

References

[Gi86] Walter Gilbert, "Origin of life: The RNA world," ***Nature 319***, 618 (1986).

[La04] A. Lazcano; J. L. Bada "The 1953 Stanley L. Miller Experiment: Fifty Years of Prebiotic Organic Chemistry", ***Origins of Life and Evolution of Biospheres 33*** (3), 235-242 (2004).

[Le03] Harold Levin, ***The Earth Through Time, seventh Edition,*** Wiley (2003).

[Mi59] Stanley L. Miller and Harold C. Urey "Organic Compound Synthesis on the Primitive Earth", ***Science 130*** (3370), 245-251 (1959).

[Mo24] E. R. R. Moody, et al, "The nature of the last universal common ancestor and its impact on the early Earth system," ***Nature Ecology & Evolution 8***, 1654–1666 (2024); arXiv:2412.14265v1.

[Ne13] M. Neveu, H.-J. Kim, and S. A. Benner, "The 'Strong' RNA World Hypothesis: Fifty Years Old", ***Astrobiology 13*** (4) 391-403 (2013).

[Pa21] Klaus Paschek, et al, "Meteorites and the RNA world II: Synthesis of Nucleobases in Carbonaceous Planetesimals and the Role of Initial Volatile Content", preprint arXiv:2112.09160v1 [astro-ph.EP] (2021)

[Ri62] Alex Rich, "On the problems of evolution and biochemical information transfer", ***Horizons in Biochemistry***, edited by M. Kasha and B. Pullman, Academic Press, New York, 103-126 (1962).

[Wa00] Peter D. Ward and Donald Brownlee, Rare Earth: Why Complex Life is Uncommon in the Universe, Copernicus, New York (2000), ISBN: 0-387-98701-0

3

Onward and Downward: Aging Theories, Philosophies, and Fallacies

3.1 What Is Human Aging?

Here are some questions for which modern bio-medicine can provide no informed consensus answers:

- *Exactly what is human aging?*
- *Why and how does it happen?*
- *Can it be slowed, halted, or reversed by biomedical interventions?*

The root cause of and mechanism behind human aging has been one of the deep mysteries of contemporary bio-medicine. Presently, there is no informed consensus on a solution. However, there has been no shortage of variant theories and speculations about human aging that have sought to fill this gap. Some of these we will consider here, focusing on published theories.

3.1.1 Theories of Aging, Speculations, and Diversions

Over the years, it has been suggested that the cause of aging is: (1) the accumulated wear and tear on cells [Cu63], (2) the accumulated genetic mutations to nuclear DNA and the correlated decline of DNA repair mechanisms [Me52, Yo21], (3) the accumulated damage of oxidizing free radicals [Ha56, He11], (4) the progressive shortening of telomeres [Fo96, Ol73], (5) the progressive decline of the immune system [Wa64], (6) the progressive accumulation of "zombie" senescent cells [Is00, Pe11], (7) the progressive loss of stored information [Lu23], or (8) the presence of an intrinsic "aging

J. G. Cramer, *How to Live Much Longer*, Copernicus Books, https://doi.org/10.1007/978-3-032-17741-4_3

clock" that counts down our epigenetic programming [Wa22], enforcing the transition from young to old and opening the door to the diseases of old age. These are only the major theories of aging, and there are also many others that I chose not to list. There is perhaps a kernel of truth in each of these ideas, but none has succeeded in becoming the accepted mainstream theory of aging or has been uniquely identified as the root cause of all the others.

Among these divergent theories of aging, the one of particular interest is item (3), the *free radical theory of aging*. Although flawed in its basic mechanism, it is actually rather close to the process that we now identify here as the probable root cause of human aging. The concept that reactive oxygen species (ROS) damage exerts a fundamental influence on aging was first proposed in 1956, when Denham Harman (1916–2014), then at U. C. Berkeley, introduced the "free radical theory of aging" [Ha56]. At the time, the roles of mitochondria and mtDNA in cell energy were not very well understood, so Harman was not very specific about just what parts of the cell were actually being damaged by oxidation.

By 1972, the critical role of mitochondria in cell operation had come more into focus in the biological community, and so Harman, now a professor at U. Nebraska, was motivated to refine his theory, that became the "*mitochondrial* free radical theory of aging" [Ha72]. In the new version of the theory, he proposed that aging results from the accumulation of damage to critical biomolecules, specifically affecting mitochondria. That damage was primarily attributed to the excessive production of highly toxic ROS within mitochondria. It was suggested that there was a positive feedback loop in which ROS damages mitochondria and damaged mitochondria produce more ROS. Support for the free radical theory stemmed from observations that mitochondrial DNA (mtDNA) had a significantly higher mutation rate, presently estimated to be around 20 times greater than that of nuclear DNA, and possessed less efficient replication and repair machinery than that of nuclear DNA.

In the closing decades of the twentieth century, the general public began to learn about and accept Harman's free radical theory. This created a bonanza for the health supplement industry. Sales of antioxidant supplements went through the roof. In the popular press, the health news triumphantly proclaimed that the secret for slowing human aging had been discovered and that we could fix many of the problems of aging by simply including antioxidants in our supplements and diets. Interestingly, that was decades ago, and the free radical theory has been shown to be a gross oversimplification. Nevertheless, some of today's "health experts" are still riding Harman's antioxidant horse.

After this initial burst of enthusiasm, it became increasingly clear that the mitochondrial free radical theory had serious problems [He11]. Its central tenets concerning the need for antioxidant supplements, the suggestion of direct ROS-driven mtDNA damage, and the "vicious cycle" (where mtDNA damage leads to increased ROS, which makes more mtDNA damage in a positive feedback loop), have all faced increasing challenges and motivated extensive revision of the theory. Contemporary research indicated that taking antioxidant supplements had little effect, that mitochondria, even under tightly controlled *in vitro* conditions, produce substantially less ROS than Harman had estimated, and that experimentally induced ROS-type oxidative stress only results in rather low levels of mtDNA damage.

Further, in 1979, the prominent British scientist and physician Alex Comfort (1920–2000), already famous for his 1972 best-selling book ***The Joy of Sex***, wrote a book [Co79] containing a scathing critique of Harman's mitochondrial free radical theory and expressing strong doubts that damage to mitochondria could possibly play any significant role in human aging. At that time, it had become known that mitochondria were constantly being damaged, cleared by mitophagy, and rapidly restored by mtDNA replication and mitochondrial fission in a frantic few-week cycle, even in non-dividing cells. Based on this new knowledge, Comfort argued that mutations in mtDNA could not possibly accumulate enough damage to drive aging, because any mutated mitochondrion would soon be removed through mitophagy and replaced by the vigorous replication of the healthy ones.

In the following decade, subsequent research showed that Comfort's plausible conclusions were incorrect. Mitochondrial DNA mutations *do* accumulate progressively with age, and they are at least one of the major causes of human aging. The inability of cellular cleanup mechanisms to completely remove damaged mitochondria, the poor mitochondrial repair mechanisms, and the rapid exponential growth of deletion mutations [Va23] in older humans all lead to a devastating mtDNA damage build-up and a resulting ATP energy shortage. Nevertheless, the publicity damage to Harman's mitochondrial free radical theory was already done, and the idea never regained its former prominence in medicine and biology as a leading theory of aging.

The mechanisms by which mtDNA is progressively damaged by accumulating three distinct forms of mutation, and why these different mutations occur and grow at different functional rates, are far more complicated than Harman's simple "damage by ROS" theory. For interested readers, the details of the more accurate mtDNA damage progression are presented in Appendix B below.

3.1.2 Wiping the Slate: Human Reproduction and the Yamanaka Factors

Nature has already provided us with a very strong hint that, at least at the cellular level, *age is indeed reversible.* We know that such cellular age reversal is possible because it clearly occurs during human reproduction (or else we would not be around to worry about it). When a man and a woman, perhaps age 30, conceive a child, both parents contribute their 30-year-old 23 chromosomes of nuclear DNA to the embryo, and the mother also starts the process with one of her oocyte reproductive cells filled with her mitochondria and their mtDNA. However, while the oocyte is one of her own cells, it has had all of her accumulated environmental damage and mtDNA mutations **somehow erased**. Further, the 30 years of epigenetic programming of all of the parent's nuclear DNA has been reset to zero. The new embryo somehow starts life with brand new embryonic cells, not the mother's 30-year-old ones (see Appendix C below).

How is this arranged? There is now a body of research providing a plausible answer to this question: Nature has arranged for a special set of reset proteins that perform the epigenetic clock "rewind" task. Takahashi and Yamanaka in 2006 reported the discovery of what are now called the "Yamanaka Factors", four reset proteins designated by the labels OCT4, SOX2, KLF4, and c-MYC (abbreviated "OSKM") that are actively produced in the cells of embryos [Ta06]. When mouse versions of these proteins were added to laboratory cultures of several cell types taken from older mice, the old cells were transformed into seemingly brand-new ones, pluripotent stem cells able to transform into many different young cell types. As a follow-up, the Sinclair Group at Harvard reported that aging ***human*** cells could be similarly transformed by adding OSK to human somatic cell cultures to accomplish the same epigenetic reset task [Yu07]. The group did not include the c-MYC (M) protein out of caution, because it has an association with tumor formation.

The Sarkar group at Stanford in 2020 reported [Sa20] that they had used an "OSKMLN cocktail" of reset proteins (where L is LIN28 and N is NANOG). These were delivered by synthesized messenger RNA (mRNA) to cultures of a wide variety of aging human cell types for 12 days, after which the cells had all become young pluripotent stem cells. They also found that the same cell cultures could be reset to a younger state while still retaining their original functional cell type differences and avoiding the pluripotent state if the treatment was halted after 4 days. In other words, the Sarkar group demonstrated that human cells can indeed be rejuvenated while retaining their functions. There is now a demonstrated procedure for "wiping the slate" of aging human cells and restoring them to their previous younger condition without damaging their ability to perform their assigned task in the human body.

There are three possible problems with this approach to cell rejuvenation, however: (1) the 4 day exposure trick is not universal in duration, and different cell types are likely to require different times of exposure to the transforming mRNA to do an epigenetic reset without losing their functional settings; (2) the needed targeting and carefully timed delivery of the mRNA to specific cell types is a difficult and only partially solved problem; and (3) the cellular reprogramming requires considerable ATP energy, and if that energy is not available because of age-related mitochondrial degeneration, the process may fail.

Very recently, an alternative epigenetic reset protein has appeared on the scientific horizon. A 2025 paper [Ca25], with authors mainly associated with the UK bio-startup ***Shift Biosciences,*** reported the discovery of a previously-unknown human gene that, using the company initials, they gave the name "SB000". This mysterious gene, not otherwise identified, produces a reset protein that, in experiments with human skin (fibroblast) cell cultures, does a remarkable job of resetting the cells' epigenetic profile to a much younger profile *without* causing the cells to lose their somatic identity or to show any evidence of pluripotency. SB000 appears to be the rough equivalent of OSK, but with a built-in automatic stop before the cell forgets its normal function. This work implies that an epigenetic reset can be readily accomplished while avoiding the OSK targeting and timing issues discussed above. It appears to be a human protein that could be freely administered to reset the epigenetics of any cells in the human body to a younger epigenetic profile. If valid, SB000 is a very important epigenetic reset breakthrough.

There is a remaining problem, listed above. with both the OSK and the SB000 epigenetic reset strategies. If applied to *very old humans*, these reset techniques are likely to have a problem with the *available ATP energy*. Epigenetic reprogramming requires an estimated 3–10 times more ATP energy than the normal operation of a typical somatic cell. Therefore, very old cells that are in the throes of a severe ATP energy crisis are likely to lack the energy resources needed to complete the reset job, perhaps halting the reset process partway through. We do not know what happens to cells that have an epigenetic reset halted in midstream, but the consequences are likely to be undesirable and damaging.

3.1.3 Mothers and Their Mitochondria

In human reproduction, the mitochondria and mtDNA of the new child come exclusively from its mother. The mother's original reproductive cells, precursors to the oocyte, initially contain a relatively small number of mitochondria (perhaps fewer than 10–100 in the very early stages). As these cells become oocyte eggs during fetal development, the mitochondria multiply until a very large number (as many as 600,000) of them are churning out the large quantity of ATP energy required for early embryonic development. In this transition, the mitochondria are parceled out among the dividing cells. In this process, a mtDNA *winnowing* happens. It is a complicated set of events described in some detail in Appendix C below.

The winnowing has the net effect that all the mitochondria in the developing embryo have mtDNA that are copied from only about 7–10 of what are normally the mother's healthiest and most robust mitochondria. These select few mtDNA are then rapidly replicated until there are around 1,000 mitochondria in each developing cell, each with many mtDNA copies. This mitochondrial "bottleneck" process normally ensures with high probability that only the most robust and healthy of the mother's mitochondria are passed on to the developing embryo. This is how Nature "clears the slate" and starts the developing child with essentially "new" mitochondria containing undamaged mtDNA.

It should be mentioned, however, that occasionally, by random chance with small odds, one or more of the mother's damaged or mutated mitochondria may slip past the bottleneck and be amplified within the embryo. That only happens perhaps once in 5,000 or more births, but it leads to miscarriages and to children born with debilitating or fatal mitochondrial genetic diseases like MELAS. (Note: mitochondrial transplantation, to be discussed later, should offer an effective treatment for such genetic disorders.)

3.1.4 Of Mice, Pigs, and Men: Aging Drivers and Transplants

The standard laboratory mouse has been used in a large fraction of all longevity experiments. They have an average lifetime of about 26–30 months and a maximum lifetime of about 3 years. Their short lifespan enables researchers to perform tests on them and to publish the results in a few years, which perhaps accounts for their popularity among longevity experiments.

The age-related effect that sets the hard limit on maximum lifetime in mice is not known, but there are suggestions that it is related to the high probability of tumor formation in older mice. In any case, the maximum lifetime limit for short-lived mice (~ 3 years) very likely has a different root cause from that limit for long-lived humans (~ 120 years). It is becoming clear that in humans, the maximum lifetime limit is likely to be set by accumulated and accelerating mitochondrial damage. In mice, their lifespan is too short for very much mitochondrial damage to accumulate, so other factors are likely to be in operation that set their lifespan limit.

However, there have been innovations from genetic engineering that bring the mtDNA state of aging mice much closer to that of aging humans. In 2004, a group in Finland and Sweden [Tr04] used genetic engineering involving the knock-in of a particular nuclear gene to produce the D257A mutation in a strain of mice, causing them to have a *proof-reading deficiency* in their PolgA replication protein, which is the nuclear DNA encoded catalytic subunit of mtDNA polymerase Pol γ. This mutation causes the process of mtDNA replication in these mice to be much more error-prone, so that *they accumulate mtDNA point mutations at a much higher rate than normal laboratory mice*. These mtDNA "mutator mice" have a normal appearance until about the age of 25 weeks, after which they begin to show effects of rapid aging, spinal curvature, and progeria. Their median lifespan is about 48 weeks, and all of them die before the age of 61 weeks (about a year and 2 months), with about 39% of the lifespans of standard lab mice.

In 2005, another group in Finland used genetic engineering to induce a mouse mutation [Ty05] that in humans is attributed to a mitochondrial defect in the Twinkle enzyme and that causes weakness/paralysis of eye movement muscles. They found that multiple *mtDNA deletions* accumulated in the tissues of these "Twinkle mice". The mice showed early signs of insufficient energy production, primarily affecting muscles, at the age of 52 weeks, but they did not show other signs of accelerated aging and had about the same lifespans as standard lab mice.

In 2007, a group mainly at the University of Washington [Ve07] compared the mutation rates and lifespans of standard laboratory mice, mice with genes that had both copies (2C) of the PolgA mutation described above, and mice that had only one copy (1C) of the PolgA mutation and developed mtDNA mutations at about 30% of the 2C rate. They found that 1C mice had essentially the same lifespan as standard mice, while 2C mice had the shorter lifespan described above. They also found from precision DNA sequencing that the mutation frequency in standard mouse mtDNA is more than ten times lower than previously reported. Their 1C mice had 11 times more point mutations than normal without showing aging effects, and one of them had 500 times more mutations without obvious aging. They concluded that *accumulated point mutations in mtDNA do not limit the lifespan of standard lab mice*. That is consistent with evidence that mouse mortality is highly correlated with tumor development.

In 2008, the same group continued their investigation of the effects on mtDNA damage in mice having one and two copies of the PolgA mutation [Ve08], but they focused on the production of mtDNA *deletion* mutations. They found that the PolgA mutation not only produces point mutations, but also produces frequent *deletion* mutations, particularly when replicating mtDNA containing double-strand breaks. They concluded that *DNA deletions and clonal mutations drive premature aging in mitochondrial mutator mice*. In other words, while mtDNA *point* mutations do not limit mouse lifetime, mtDNA *deletion* mutations do, at least for the short-lived homozygous mutator mice in their study.

Thus, these two mutant mouse strains allow experimenters to investigate the effects of each of the two kinds of mtDNA damage that accumulate in aging humans, point mutations and deletion mutations. They found that the order in which the symptoms of aging appear in humans was more accurately reproduced in the Twinkle mutator mice than in the PolgA mutator mice, implying that the former are a better model for human aging. Further, it supports the hypothesis, discussed in detail in Appendix B below, that accumulated mtDNA *deletion* mutations (not point mutations) are the primary driver of aging in very old humans [We25c].

Is there a connection between mtDNA mutations and epigenetic programming? Direct experimental evidence, particularly from studies on these mutator mice and from *in vitro* cell culture studies, demonstrates that mitochondrial dysfunction, including reduced ATP production induces changes in nuclear DNA methylation profiles and chromatin structure. Such changes include histone H3K9 di-methylation marks, traditionally associated with gene silencing, in nuclear DNA. Furthermore, a quantifiable link has been established between the functional impact of mtDNA changes and an older epigenetic age in young adulthood. In other words, these mutator mouse studies provide a definite example in which mtDNA damage causes age-related changes in epigenetic programming.

In 2018, a group at the University of Alabama [Si18] used genetic engineering to create a strain of mtDNA-depleter mice in which mutation-modified Pol γ polymerase could be switched on by the presence of the antibiotic tetracycline. This had the effect of inducing reduction in the number of mtDNA copies in the whole animal on command. They were then able to observe the effect on the mouse physiology of depleting its mtDNA at age 8 weeks and then restoring the mtDNA after two months of depletion.

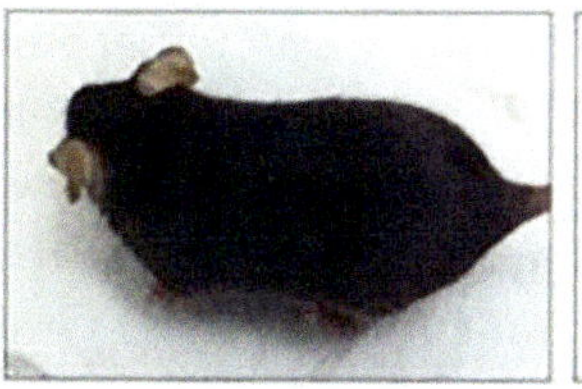  

Fig. 3.1 A depleter mouse before, after mtDNA depletion, and after restoration. *Credit* Reference [Si18] and U. Alabama, used with permission)

The experiment demonstrated an amazing onset of the symptoms of aging, followed by the removal of those symptoms when mtDNA was restored. This work strongly supports the hypothesis that progressive damage to mtDNA is a principal cause of human aging [We25c] (Fig. 3.1).

In 2015, Dr. James McCully and Dr. Sitaram Emani of the Boston Children's Hospital pioneered the use of mitochondrial transplantation to treat newborn humans with severely damaged hearts. McCully had previously tried mitochondrial transplantation on pigs with heart damage. He would take a plug of muscle tissue the size of a pencil eraser, extract about 20 billion mitochondria from it, and inject these mitochondria directly into pigs' damaged hearts. To his surprise, the mitochondria "moved like little magnets" to the needed places and began supplying energy. The pig hearts recovered.

Emani and McCully, using similar procedure [Gu21], performed the first human mitochondrial transplant on an infant girl named Georgia, whose heart had been injured during surgery and was being kept alive with a heart-lung machine. Dr. McCully extracted about 1 billion mitochondria from a tiny sample of the infant's abdominal muscles, and Dr. Emani injected these into the most damaged areas of the infant's failing heart. They reported that they could see the damaged areas of the infant's heart turn from gray to pink and revive as the mitochondria arrived. Remarkably, the heart of the baby began beating normally within two days.

3.1.5 Evolution and Aging in Social Species

In considering evolutionary effects, it is sometimes confidently stated by some "expert" that the natural selection process stops after reproduction, so that post-reproductive humans can be viewed as "on their own", without the support of evolutionary processes for developing improvement of their biology. In my view, this simplistic conclusion is incorrect.

As the great sociobiologist E. O. Wilson has taught us [Wi12], that in viewing evolution, it is important to realize that humans, along with ants, termites, bees, and wasps, have made the great evolutionary leap to become a *social species.* Because of this, all of these social species have gone on to dominate their respective ecological niches. As evidence of their success, the social insects represent over 60% of terrestrial insect biomass. In all social species, the basic element of evolution is not the *individual* but the *group.* Group fitness is the evolutionary driver, and it brings fitness to the group members, whether they are young or old. All are brought along on the evolutionary progress ride.

There are mathematical simulations of evolutionary dynamics [Ha64] that demonstrate the importance of group selection in the evolution of social species. For humans, the group includes infants, young, breeding adults, and post-reproductive adults. The forces of evolutionary selection affect all of these components, all of which contribute to group survival.

This is the likely reason why we humans are the longest-lived of the primates, with a maximum lifespan of around 120 years. Our evolutionary cousins, the gorillas, have a maximum lifespan of about 60 years, and the chimpanzees have a maximum lifespan of about 80 years.

In a group evolutionary sense, the old humans are needed to preserve shared group learning and knowledge, to educate and protect the young, and to support the young and the breeders with their skills, wisdom, and accumulated resources. Evolution has been pushing hard to make humans live longer, better to support the group, but this has hit the limits imposed by the inevitable shortage of ATP energy arising from progressively accumulated mtDNA damage, a consequence of the Faustian Bargain described in Chap. 2 above.

3.1.6 Is it the Entropy or the Energy?

Some longevity entrepreneurs have argued that the modest anti-aging interventions and products that they are selling are "all there is or ever will be" available, because the reversal of human age is physically impossible, as a consequence of physics and the laws of thermodynamics. They assert that aging is just a manifestation of increasing *entropy*, a quantification of disorder, which must always increase with time, as required by the 2nd Law of Thermodynamics.

As a physicist with some experience in actually calculating and dealing in detail with entropy (in ultra-relativistic heavy ion collisions at CERN and Brookhaven), I can confidently say that such assertions are absolute nonsense. The 2nd Law applies only to *closed* physical systems, in which no energy or matter flows in or out across the system boundaries. That hardly describes a human, aged or otherwise, who is breathing, drinking, eating, and excreting waste, and is clearly raising the entropy of the food consumed in order to lower his own. Many of the touted interventions put new material, e.g., young plasma, into the system. Entropy is an overall disorder bookkeeping quantity, best applied to statistically large systems of identical gas molecules or fundamental particles. In open systems, dealing with entropy is much more difficult, and perhaps the more meaningful quantity is the entropy *flow*.

That said, in 2007, a group of engineers at Texas A&M University [Si07] did attempt to calculate the entropy and its flow for humans as they aged. Interestingly, they found that entropy generation in infants is three times higher than in the elderly. From the perspective of 2025, however, while this two-decade-old work on entropy is interesting as a demonstration that such analysis is possible, it has not sparked new research or proved to be particularly useful as a basis for the ongoing investigation of human aging in the biomedical literature.

On the other hand, *energy* is a powerful concept for analyzing and understanding biological processes including aging. The omnipresent ATP molecule, the cellular energy source, essentially stores its energy like a compressed spring, "cocking" one of its phosphate components to a high-potential-energy state, which, when released, delivers about 0.33 electron-volts of energy. (For comparison, it requires 5.14 electron-volts of energy to ionize sodium, and to break a typical C–H bond requires over 4 electron-volts.) This ATP energy is released when the energized phosphate is ejected like a bullet from a gun, and the ATP (*adenosine tri-phosphate*) becomes ADP (*adenosine di-phosphate*), ending with two phosphates instead of three. In respirating cells, after this ATP energy release, the "waste product" ADP molecules are collected, returned to the mitochondria and recycled by forcibly adding another phosphate to produce more new ATP. The healthy human body typically produces, uses, and recycles about 1.5 times its body weight in ATP over the course of one day.

It is relatively easy to calculate the amount of ATP required for a specific cellular process. This allows us to estimate the net energy cost of a particular overall cellular operation in units of the ATP molecules. For example, the daily operation of a typical human somatic (body) cell requires about 10^9 ATP units (but more active neurons and muscle cells can require 10^{10} ATP units or more), an apoptosis to clear one damaged cell requires about 10^6–10^7 ATP units, the production of a single DNA or RNA nucleotide requires about 50 ATP units, the replication of one mtDNA ring requires approximately 9×10^4 ATP units, the replication of one complete mitochondrion requires about 7×10^9 ATP units, and the epigenetic reprogramming of one cell requires several days and about 10^{10}–10^{11} ATP units or more. Knowledge of such energy requirements, as discussed further in Chap. 8, is essential in understanding how the human body behaves as it adapts to an energy shortage situation like old age.

3.2 Who Ages Well?

About a decade ago, two of my former high school classmates sent me a.pdf copy of our old Class of 1953 pre-graduation group photograph from Lamar High School in Houston. The picture had been modified by adding small crosses at chest level, superimposed on all of our classmates who had died since the picture was taken. There was a dismayingly large number of these crosses. I noticed that the deceased seemed to be the classmates I remembered as being taller, more mature, more athletic, and higher on the social ladder than I had been. It seemed that it was my more slowly-maturing, not very athletic, and not very socially active classmates who had survived into our 80s. Perhaps, there is a message here: *early maturity has a cost, while slow and steady wins the race.*

So what allows some of us to push the upper end of the human aging envelope while others cannot? My mother, her father, and her maternal uncle all survived into their 90s. Clearly, it helps to have good genes, to inherit from your ancestors no genetic diseases or predispositions to cancer, diabetes, heart failure, etc. A study of the time interval between death dates for human twins found that the mean death-date difference in *identical* twins was 6.65 ± 5.6 years, while in *fraternal* twins it was 8.66 ± 7.2 years. This indicates that identical twins with the same genes tend to have closer death events than fraternal twins having somewhat different genes, but the net effect is only a couple of years. Death does not seem to be tightly programmed by the genome, and factors beyond genetics must be important.

Environmental factors are certainty important. One should avoid living and working in regions with heavy pollution, workplace radiation or chemical vapors, contaminated water, many mosquitoes or other disease-vector insects, high communicable disease rates, high radon from underground, or substandard medical care.

It is also frequently advised that maintaining good health in old age requires regular exercise of the skeletal muscles. Exercising the mind through working puzzles, reading unfamiliar material, and encountering new ideas is also advised. These practices are certainly beneficial, but it is not clear what the size of their effect is on the overall well-being of the aged.

And of course, the other prominent longevity factor is *lifestyle choice*. If one chooses to smoke, to drink alcohol in excess, to do drugs, etc., those choices will certainly diminish human longevity. In clinical studies involving large populations, researchers are always careful to divide the group into smokers and nonsmokers, because the negative effects of smoking on health and longevity are so large. Similarly, occupations like coal and uranium mining, commercial fishing, logging, manufacturing involving chemical vapors, long-haul trucking, jobs with radiation exposure, and shift work involving rotating and changing sleep periods are known to have negative effects on health and longevity. Longevity isn't just about physical danger. Stress, poor work-to-life balance, and exposure to harmful substances all play a role.

Finally, for the goal of aging well, some interventions, supplements, and treatments are available that are reputed to be beneficial. In Chaps. 7 and 8 below, we will discuss and evaluate many of these interventions. I have been using some of them for a number of years, which may be related to the fact that I am still in good health at age 91 and here to write this book.

3.3 Is There a Maximum Human Age?

As discussed above, we humans, with our apparent maximum age limit of around 120 years, are the longest-lived of the primates. However, among all vertebrate species on Earth, our limit is far from the maximum-age record holder. The *saltwater crocodile* matches humans with a maximum age of around 120 years. The *Seychelles giant tortoise* has a maximum age of 190+ years, and the *bowhead whale* has a maximum age of 200+ years, as does the *rougheye rockfish*. The *Greenland shark* has perhaps the highest maximum age of vertebrates, which some estimates place as high as 512 years. Invertebrate adult *T. dohrnii* jellyfish under stress can revert to their post-larval stage and are said to be "immortal" and without a maximum age limit.

While the mechanism that allows these species to have such very long maximum ages is not understood, it clearly helps to be aquatic, to live in cold water, and to have a relatively slow metabolism. Recent gene sequencing analysis of some of these species indicates the genomic presence of many more copies of those genes related to DNA repair and replication than is the case for humans. It would be very interesting to know more about their mtDNA replication and repair mechanisms, and whether they are better than ours.

As of this writing, the record for the oldest human is still held by the late Jeanne Calment (1875–1997) of Arles, France, who died at the age of 122 years 164 days. The runner-up is Kane Tanaka (1903–2022) of Japan, who died at the age of 119 years 107 days. The oldest human still living today is Ethel May Caterham (1909–) of Shipton Bellinger, Hampshire, England, who was 116 years old last August, 2025. The oldest living male human is João Marinho Neto (1912–) a Brazilian farmer who was 113 years old on October 5, 2025. There are also many unverified reports and folk tales of much longer-lived humans. We note that human average and maximum lifespans for males and females are markedly different, with males dying earlier. There is some evidence that this difference is attributable to male/female differences in the expression of the mitrochondria-related genes in the cell nucleus, which leads to sex-determined differences in mitochondrial health and respiration.

When humans live to such advanced ages, their ultimate causes of death differ somewhat from those of the "younger" elderly population. While traditional age-related diseases like cancer, heart disease, and stroke are still present in the extremely old, they often appear much later and sometimes are avoided altogether. Instead, the cause of death for the extremely old is often described as a "general exhaustion of organ reserve" or "senility." This means that their bodies have simply reached a point where multiple organ systems gradually decline and lose their ability to function or to recover from disruptions. It's less about a single specific disease overwhelming the body and more about a systemic overall breakdown due to extreme aging. Another leading cause of death in the extremely old is *senile systemic amyloidosis*, involving a protein called *Transthyretin* misfolding and depositing in blood vessels, leading to issues like congestive heart failure and to neural diseases like Alzheimer's, Parkinson's, and Huntington's.

There have been aging studies [Sg14] comparing fibroblast (skin) cell cultures taken from 27, 75, and 100-year-old human subjects to examine the mitochondrial structure vs. age. They found that on the oldest samples, the mitochondria had joined together to form a *hyperfused mitochondrial network*. Typically, a single healthy mitochondrion contains over 10,000 Complex V synthase sites, each producing around 200 ATP molecules per second. However, the hyperfused mitochondrial network has an effective surface area of inner membranes and cristae that is considerably larger (by ~ 2–3) than the sum of the inner surface areas of the pre-fused mitochondria, so that the total number of Complex V sites has been significantly increased in this configuration. The researchers suggest that this change found in the oldest cells serve the purpose of resisting unwanted mitophagy and increasing the efficiency of ATP production per mitochondrion in response to the age-induced cellular energy shortage. This rearrangement of mitochondria may have short-term energy benefits, but in the long term, it will interfere with mitochondrial replication and renewal by fission and will contribute to age-related decline.

If we view these maximum-age-limiting show stoppers from the perspective of cell energy, an underlying cause is clearly suggested. The individuals who manage to avoid the common age-related fatal diseases and approach the 120-year limit, because of such long exposure times, must always have severely damaged mtDNA and barely functioning mitochondrial ATP energy production. Cell repair, replication, and replacement require considerable energy, and the reported "organ exhaustion" is likely a consequence of this severe energy shortage that blocks all cell repair and replacement. Repair of accumulating v nuclear DNA damage, which is very effective but has a high energy cost, will be "starved" to inaction by this energy shortage.

As for amyloidosis, the healthy human body has elaborate bio-machinery for dealing with misfolded proteins. It involves repair by specialized chaperone proteins, or their removal by the proteasome system, autophagy, or the cell's endoplasmic reticulum. However, all of these systems require significant ATP energy to function. If such energy is not available, the systems cannot function, and inevitably, misfolded proteins like Transthyretin will build up.

In the coming decade, we should finally be able to treat a few of these extremely old humans with effective high-volume mitochondrial transplantation [Be23]. This should restore their ATP levels to a healthy state and dispel their body's severe energy shortage. It will be very interesting to observe the effect of such interventions in the extremely old. The results should be quite spectacular.

In speculative fiction and sensationalist media headlines, the term "immortality" is often invoked and conflated with any extension in human longevity. Such use is misleading and should be discouraged. We are now in the foothills of the mountain range of extensions of the human lifespan, and correcting mitochondrial damage and decline is only the first real mountain to be summited. In the unexplored territory behind it are likely to be many more such mountains.

Even if mitochondria are fully restored and a full supply of ATP energy becomes available, there will undoubtedly be unknown and potentially lethal agents building up and vital enzymes essential to life falling off, even in an energy-restored body. If we fix all of the known age-limiting diseases, there will *still* be another hard limit on the human lifespan. Even with healthy mitochondria, youthful epigenetic programming, proper-length telomeres, and no senescent cells, there will remain some hidden processes that will eventually place limits on the maximum human age.

One of these limiting factors is our *nuclear DNA*. Even if it is well protected and equipped with very effective repair mechanisms, it is nevertheless continually sustaining unrepaired damage from many sources. One of these is damage from ionizing radiation from the radioactive decay within the body itself of the unstable natural isotopes ^{3}H, ^{14}C, ^{40}K, and ^{131}I, all present in the molecular structure of the cells, and alpha-particle damage from inhaled ^{222}Rn inert gas coming up from underground radioactive minerals. There is also continuing damage to the genome from cosmic ray muons raining down from the stratosphere and from mutation-inducing environmental chemical toxins.

Further, it has recently been discovered [Pi25] that fragments of mtDNA somehow find their way into cell nuclei and can insert themselves into the nuclear DNA with damaging effects, particularly to brain tissue. Although nuclear DNA accumulates mutation damage at a rate about 20 times slower than mtDNA, its mutational decline will ultimately reach the critical point of becoming life-threatening and life-limiting.

Another consequence of extreme age is the decline of the extracellular matrix. The human ELN gene that codes for the structural protein *elastin* is silenced when we are relatively young but have stopped growing. This leads to the gradual degradation of the structural integrity of the body's extracellular matrix, which is made from elastin and holds everything together. For example, the sag in women's breasts with increasing age is a consequence of elastin loss. If the onset of renewed energy does not reactivate the ELN gene, and other interventions do not correct the problem, the aging body will eventually lose its structural integrity and literally sag and fall apart.

There are very likely to be more of these lurking "time bombs" associated with increasing age that will only become apparent when we have lived long enough for them to appear. Thus, the future of human longevity promises to be a horse race between a series of emerging age-related problems and an offsetting series of new interventions and treatments invented to deal with them as they appear.

In the area of extreme age, there is one unresolved question: *How do the extremely long-lived vertebrates listed above manage to live for so much longer than the oldest humans?* The answer is not clear, given that their own mtDNA should be accumulating debilitating damage at about the same rate as ours, which should be creating the same growing shortage of ATP energy.

From the first DNA sequencing studies on bowhead whales and a few others, it has been found that genomes contain multiple copies of certain genes for repairing nuclear DNA. Soon, we should be able to study these total genomes and identify longevity genes in much more detail. When this is accomplished, I predict that we will find that these long-lived vertebrates also have evolved some extra mechanisms for mtDNA repair, or have better proof-reading during mtDNA replication, or have better mitophagy, or perhaps they have evolved some other unknown mechanism for renewing their mtDNA on a more regular basis. As we scientists always say, *there is a great need for more research in this area.*

3.4 Philosophies of Aging: Accept it or Fix it?

It is interesting to view the wide variety of attitudes of the general populace on one central aging question: is human aging is an unchangeable natural condition, or is it one of the many bio-technical problems that can be analyzed, changed, and corrected. Whenever the ***New York Times*** or the ***Washington Post*** runs an article reporting recent progress in understanding, slowing, or reversing human aging, the reader comments are quite telling. Over half of the comments are usually something like "*Aging was put in place by the Will of God, and you scientists are misguided fools committing a hideous sacrilege to think you can change that,*" or "*Aging is a natural process of life, and from an ethical and moral standpoint, we must learn to accept and accommodate it rather than fighting it,*" or "*Get over it, Suckers! We're all gonna die!*" Other commenters express a strong preference for "getting it over with quickly" rather than wasting away in a nursing home while bankrupting one's family. One gets the impression that over half of the US population, or at least half of the news readers and commenters, are prepared to suffer the degeneration, dementia, and decline of old age or early death rather than entertain the idea of doing anything about it.

Therefore, a key question surfaces: *Is it naive to think that human aging can be halted or reversed*? I would first say that the question absolutely *not* naive. As a physicist, I view biological systems only as rather more intricate and complicated versions of the physical systems with which I have been dealing for five or six decades. So the answers are obvious. *Of course,* there is at least one root cause of human aging. *Of course,* there must be ways that human aging can be slowed, stopped, or even reversed. We simply need more knowledge and understanding, so that we can identify and implement such measures. In that effort, with our new understanding of the key role of mitochondria in aging, we may already have crossed an important understanding threshold and may be well on our way to implementing effective treatments for reversing human aging.

Some of us in the longevity community, in discussing anti-aging interventions with our friends and acquaintances, have noticed that there is a definite progressive shift in attitudes and philosophies about human aging as we grow older. An age of 50 is perhaps the threshold when the feeling begins to wear off that you are too young to worry about growing old. At age 60, aging is a definite worry, and you are beginning to lose old friends and relatives, but the prospect of being very old and infirm still seems quite far away. At age 70, the reality of aging sets in. Your hair is graying and thinning, your muscles are losing their former strength, many friends your age are now gone, and you notice that you are attending more funerals than weddings. At age 80, aging is a looming presence, the major medical events in your life are growing more frequent, and your children are watching you closely and telling you about acquaintances who are very happy in retirement and nursing homes. At age 90, you have lost almost all of your former classmates and friends of youth, and even if you are in good health, you are very aware that you have only a few years left. In the last two of these decades, it is easier to convince you that longevity interventions are an excellent idea and that there *should* be an effective treatment for human aging, even if the US Food and Drug Administration thinks otherwise.

3.5 The Fallacies of Longevity and Age Reversal

It would seem that everyone has an opinion about aging. Some of the ideas are good, while some are simply wrong. In this section, we will examine some of the fallacies that have grown up about the subject.

3.5.1 Misplaced Certainty

Claims such as "everyone knows aging is inevitable" or that "life extension will cause overpopulation". They assume the assertion is true simply because it is widely believed or superficially plausible.

3.5.2 Appeal to the Natural

Suggesting that aging is "a natural process" that should not be interfered with, or that trying to extend human life is "unnatural." On the contrary, humanity has consistently sought to overcome many natural limitations, from treating infections to improving agricultural yields. The concept of what is "natural" in this context is often ill-defined and can hinder scientific progress. Nature has also given us screw-worm flies, bedbugs, smallpox, malaria, cancer, syphilis, leprosy, dementia, … "Natural" doesn't always mean "desirable".

3.5.3 "One Pill to Rule Them All"

Prof. David Sinclair, a prominent Harvard age-reversal investigator and promoter of epigenetic reprogrammming, predicted in a 2024 interview with Neil deGrasse Tyson that we may see a $10 age-reversal pill for humans within the next five to six years. He based this claim on his ongoing work with Yamanaka reset proteins. There is currently no single pill, supplement, or "miracle treatment", even in development, that has been scientifically proven to stop, slow, or reverse human aging. While some biomolecules (e.g., rapamycin, metformin, resveratrol, …) show promise in animal studies of slowing age-related symptoms and delaying mortality, their effects in humans are still largely untested, unproven, and anecdotal. As discussed elsewhere in this book, age reversal from epigenetic reprogramming has its own implementation problems associated with energy demands that may pose complications and dangers for the very old. True human age reversal may be possible. There now seem to be paths toward achieving it (see Chap. 8), but it will require massive and carefully bio-engineered interventions rather than simply popping a pill.

3.5.4 Reliance on the FDA

It is sometimes argued that no FDA-approved medication could harm mitochondria, or else it would never have been approved. Wrong! Many common drugs temporarily interfere with the operation of one or more of the mitochondrial ATP production processes. Widely-prescribed statins are known to deplete CoQ10, interfere with mitochondrial function, and elevate ROS. Moreover, antibiotics in the *fluoroquinolone* family, including the antibiotic Ciprofloxacin or Cipro (widely prescribed to treat urinary tract infections in women and as protection against anthrax), kill bacteria by attacking the replication of their loop of DNA, and do the same for mitochondria. There are examples of former US federal officials and Gulf War veterans who were given frequent doses of Cipro when traveling to anthrax-prone regions, and who are now bedridden or badly afflicted with debilitating mitochondrial failure.

3.5.5 The Sunk-Cost Fallacy

"We shouldn't invest in long-term research to investigate the causes of aging, because any payoff would take such a long time. We are already spending an enormous amount of money on the treatment of age-related diseases, and that investment is having immediate results." This ignores the potential for discoveries that partially or completely eliminate age-related diseases, removing the enormous cost of treating them and the burden they impose on the healthcare system.

3.5.6 The Distraction Fallacy

"Research on aging is just a distraction from the world's 'real problems' like poverty, homelessness, and climate change, which need far more attention and funding than they are currently receiving." Actually, it is not an either/or situation, and we would be wise to invest in *all* of these. Human aging contributes to the global disease burden and should be addressed alongside the other issues. Further, it requires only a small fraction of the funding that the other major problems would need to have similarly significant impact (Fig. 3.2).

Fig. 3.2 Detail: Lucas Cranach Sr., "The Fountain of Youth," (1546). *Credit* Public Domain

3.5.7 The "Fountain of Youth" Fallacy

Many commercial products use heavy advertising to suggest dramatic memory restoration, wrinkle removal, skin restoration, or age reversal, or claim to slow or halt aging. Current research shows that we can only slow certain specific aging processes, and this is not done with the advertised products. True human age reversal may be coming soon, but for now it is not commercially available. The "Fountain of Youth" may actually be a bio-reactor tank.

3.5.8 Driver Versus Passenger

There is some tendency, in cases where a biomarker of aging has been identified (examples: short telomeres, senescent cell buildup, ...), to assume that if this biomarker is made to change to a younger profile, the associated aging will be reversed. A distinction in aging must be made between the "driver" and the "passengers". The driver assumption is possibly true for mitochondrial damage and epigenetic programming, but it is more often wrong. The causal relationships are usually more complex, and correlational does not imply causation.

3.5.9 Misinterpretation of Animal Studies of Aging

Experimental findings in tests using model organisms like worms, flies, mice, or rats, while valuable for understanding and unraveling biological mechanisms, usually cannot be directly extrapolated to humans. What extends the lifespan of a short-lived fruit fly or worm may have no effect (or even adverse effects) on long-lived humans. The root causes of human aging are probably very different from those of short-lived species. Many interventions applied to mice, for example, frequently petting them, can produce a ~ 20% increase in lifespan. Further, there is an intrinsic research bias toward focusing on aging effects in short-lived species, because the results are available for publication much faster. Nevertheless, rigorous human trials of aging and longevity are essential, and they are fairly rare. In particular, they are often lacking for many "anti-aging" interventions, in part because the FDA discourages human testing of treatments for aging itself (as opposed to downstream diseases that are aging's consequences).

3.5.10 The "Free Ride" Fallacy

This is the assumption in cellular biology that some intervention or genetic modification will have about the same effect in all members of the same species, independent of the variable state of their supply of ATP energy among the test animal or human subjects. This ignores the obvious fact that biological processes require energy, often large amounts of energy. For example, the epigenetic reprogramming of one cell is estimated to require about 3–10 times as much energy as is required to operate one healthy somatic cell for one day. Cell division also requires about 100 times as much energy as is required to operate one healthy somatic cell for one day [Gi24]. The ongoing repair of nuclear DNA also requires substantial energy. Causing a cell to become senescent (inactive due to damage) and to stop dividing requires about 10–100 times as much energy as is needed to operate one healthy somatic cell for one day. There are few "free lunches" in the area of cell biology.

3.6 Who Is Helping; Who Is Obstructing?

Studies of human aging, its cause, and its treatment are a worldwide effort. In the past decades, despite the conspicuous absence of any direct aging-treatment funding from the US Government, there has been a dramatic rise in activity in the field of human longevity. The private nonprofit Methuselah Foundation has for two decades been funding and encouraging research and development in all aspects of aging, including investing in bio-startups like Oisin Bio, OncoSenX, Turn Bio, X-Therma, Mitrix Bio, and Nanotics.

As evidence that this is an accelerating worldwide field, recently there have been a large number of international conferences on the science of human aging, including the *Systems Aging Conference*, the *International Conference on Aging and Disease*, the *Longevity Leaders Congress*, the *Rejuvenation Startup Summit*, the B*iology of Aging Conference*, the *World Congress on Aging and Geriatrics*, and the *Longevity Conferences and Summits 2025*. Among all rapidly developing "hot" scientific fields with important new research and entrepreneurial opportunities, aging, its cause, and its treatment are near the top of the list.

At the commercial marketing level, there has also been rapid growth in the number of supplement suppliers who offer small-molecule drugs and intervention for longevity and age reversal. Some of the reputable leaders in this business are Swanson, Life Extension, Elysium Health, and ProHealth Longevity.

Unfortunately, some commercial suppliers might be considered "snake oil peddlers", engaging in heavy advertising in the media and offering questionable supplements for things like "memory improvement" and "joint restoration". Since health supplements are not regulated, and even when exaggerated claims are repeatedly made in the media, it requires years of court battles before false claims can be halted by legal actions. It is wise to consult even-handed sources like Wikipedia or Consumers' Research for advice on highly advertised aging remedies before making any expensive purchase.

On the other side of the longevity problem, ever since the 1960s, when it shifted its regulatory focus from drug "safety" to drug "efficacy", the US Food and Drug Administration has acted to raise the cost of drug progress-to-approval by a very large factor and to stifle almost all direct longevity and age-reversal research. The FDA has ruled that human aging is *not a disease warranting treatment*, because it affects *more than half of the population.* For this peculiar reason, while the CDC and FDA provide billions in funds for research and development of treatments for age-related diseases that are the downstream consequences of human aging (cancer, diabetes, kidney failure, heart disease, ...), not a penny of federal funds is spent on funding research on the causes and remedies for human aging itself.

Further, because of this "not-a-disease" rule, almost no externally funded trials testing longevity and age-reversal interventions have received approval by the FDA. One gets the impression that the FDA bureaucrats feel that human aging and death by old age are "natural processes" that should not be interfered with. This blockade has driven both valid and questionable treatments for aging offshore to escape the long reach of the FDA (Fig. 3.3).

Interestingly, this frustrating situation may be about to change. State Legislatures of states like Texas, Utah, and Montana have recently passed sweeping "Right to Try" legislation applying to cells and their products that may open the door to unfettered human testing of some new longevity and age-reversal interventions.

Fig. 3.3 The FDA and human aging. *Credit* Ratio Hartwell

References

[Be23] T. Benson, S. Patel, B. C. Albensi, V. B. Mahajan, A. Adlimoghaddam, S. Sikora, and H. Saito, "Mitochondrial transplantation and its impact on infectious disease progression: a pilot study," (2023), bioRXiv Preprint https://www.biorxiv.org/content/10.1101/2023.09.08.556161v2.

[Ca25] L. de L. Camillo, et al, "A single factor for safer cellular rejuvenation", BioRXiv preprint, https://www.biorxiv.org/content/10.1101/2025.06.05.657370v1 (2025).

[Co79] Alex Comfort, ***The Biology of Senescence (3rd ed.)***, p 81–86, Elsevier, New York (1979).

[Cu63] H. J. Curtis, "Biological mechanisms underlying the aging process," ***Science 141*** (3582), 686–694 (1963).

[Fo96] Michael Fossel, ***Reversing Human Aging***, William Morrow & Co (1996).

[Gi24] S. C. Ginther, H. Cameron, C. R. White, and D. J. Marshall, "Metabolic loads and the costs of metazoan reproduction," ***Science 384***, 763–767 (2024); https://doi.org/10.1126/science.adk6772.

[Gu21] A. Guariento, B. L. Piekarski, I. Doulami, David Blitzer, A. M. Ferraro, D. M Harrild, D. Zurakowski, J. Del Nido, J. D. McCully, and S. M. Emani, "Autologous mitochondrial transplantation for cardiogenic shock in pediatric patients following ischemia-reperfusion injury," ***J. Thorac. Cardiovasc. Surg. 162*** (3):992–1001 (2021); https://doi.org/10.1016/j.jtcvs.2020.10.151.

[Ha56] D. Harman, "Aging: A theory based on free radical and radiation chemistry," ***Journal of Gerontology 11*** (3), 298–300 (1956).

[Ha64] W. D. Hamilton, "The genetical evolution of social behaviour," I & II, ***Journal of Theoretical Biology*** *7* (1), 1–16 & 17–52 (1964).

[Ha72] D. Harman, "The biologic clock: the mitochondria?", ***Journal of the American Geriatrics Society 20*** (1972).

[He11] S. Hekimi, J. Lapointe, and Y. Wen, "Taking a 'good' look at free radicals in the aging process," ***Trends In Cell Biology 21*** (10), 569–76 (2011).

[Is00] F. Ishikawa, "Aging clock: the watchmaker's masterpiece," ***Cell Mol. Life Sci. 57*** (5), 698–704 (2000).

[Lu23] Y. R. Lu, X. Tian, and D. A. Sinclair, "The Information Theory of Aging," ***Nat. Aging. 3*** (12), 1486–1499 (2023); https://doi.org/10.1038/s43587-023-00527-6.

[Me52] Peter Brian Medawar, ***An Unsolved Problem of Biology***, H. K. Lewis & Co. Ltd., London (1952).

[Ol73] A. M. Olovnikov, "A theory of marginotomy: The incomplete copying of template margin in enzymic synthesis of polynucleotides and biological significance of the phenomenon," ***Journal of Theoretical Biology 41*** (1), 181–190 (1973).

[Pe11] Daniel S. Peeper, "Aging: Old cells under attack", ***Nature 479***, 186–187 (10 November 2011)

[Pi25] Martin Picard, "Jumping 'Numts' from Mitochondria Can Be Fast and Deadly," ***Scientific American***, January 3, 2025.

[Sa20] T. J. Sarkar, M. Quarta, S. Mukherjee, A. Colville, Paine, L. Doan, C. M. Tran, C. R. Chu, S. Horvath, L. S. Qi, N. Bhutani, T. A. Rando, and V. Sebastiano, "Transient non-integrative expression of nuclear reprogramming factors promotes multifaceted amelioration of aging in human cells," ***Nature Communications 11***, 1545 (2020).

[Sg14] G. Sgarbi, Matarrese, M. Pinti, C. Lanzarini, B. Ascione, L. Gibellini, E. Dika, A. Patrizi, C. Tommasino, M. Capri, A. Cossarizza, A. Baracca, G. Lenaz, G. Solaini, C. Franceschi, W. Malorni, and S. Salvioli, "Mitochondria hyperfusion and elevated autophagic activity are key mechanisms for cellular bioenergetic preservation in centenarians," ***Aging (Albany NY) 6*** (4), 296–310. (2014); https://doi.org/10.18632/aging.100654.

[Si07] C. Silva and K. Annamalai, "Entropy Generation and Human Aging: Lifespan Entropy and Effect of Physical Activity Level," ***Entropy 2008, 10***(2), 100–123 (2007); https://doi.org/10.3390/entropy-e10020100

[Si18] Bhupendra Singh, Trenton R. Schoeb, Prachi Bajpai, Andrzej Slominski & Keshav K. Singh, "Reversing wrinkled skin and hair loss in mice by restoring mitochondrial function," ***Cell Death & Disease* 9**, 735 (2018).

[Ta06] K. Takahashi and S. Yamanaka, "Induction of pluripotent stem cells from mouse embryonic and adult fibroblast cultures by defined factors," Cell 126, 663–676 (2006).

[Tr04] A. Trifunovic, A. Wredenberg, M. Falkenberg, J. N. Spelbrink, A. T. Rovio, C. E. Bruder, M. Bohlooly-Y, S. Gidlöf, A. Oldfors, R. Wibom, J. Törnell, H. T. Jacobs, and N-G Larsson, "Premature ageing in mice expressing defective mitochondrial DNA polymerase," ***Nature 429***, 417–423 (2004).

[Ty05] H. Tyynismaa, K. Mjosund, S. Wanrooij, and A. Suomalainen, "Mutant mitochondrial helicase Twinkle causes multiple mtDNA deletions and a late-onset mitochondrial disease in mice," ***Proc Natl Acad Sci U S A. 102*** (49):17687–17692. (2005).

[Va23] A. R. Vandiver, A. N. Hoang, A. Herbst, C. C. Lee, J. M. Aiken, D. McKenzie, M. A. Teitell, W. Timp, and J. Wanagat, "Nanopore sequencing identifies a higher frequency and expanded spectrum of mitochondrial DNA deletion mutations in human aging," ***Aging Cell 22*** (6) (2023); https://doi.org/10.1111/acel.13842

[Ve07] M. Vermulst, J. H. Bielas, G. C. Kujoth, W. C.Ladiges, S. Rabinovitch, T. A. Prolla, and Lawrence A Loeb, "Mitochondrial point mutations do not limit the natural lifespan of mice," ***Nature Genetics 39***, 540–543 (2007).

[Ve08] M. Vermulst, J. Wanagat, G. C. Kujoth, J. H. Bielas, S. Rabinovitch, T. A. Prolla, and L. A. Loeb, "DNA deletions and clonal mutations drive premature aging in mitochondrial mutator mice," ***Nature Genetics 40***, 392–394 (2008).

[Wa22] K. Wang, H. Liu, Q. Hu, *et al.*, "Epigenetic regulation of aging: implications for interventions of aging and diseases," ***Sig. Transduct. Target Ther. 7***, 374 (2022); https://doi.org/10.1038/s41392-022-01211-8.

[Wa64] R. L. Walford. "The Immunologic Theory of Aging," ***Gerontologist 4***, 195–197, (1964).

[We25c] V. Weissig, "Mitochondrial dysfunction as the "mother" of all hallmarks of aging," ***Journal of Mitochondria, Plastids and Endosymbiosis***, ***3:1***, 2560191 (2025); https://doi.org/10.1080/28347056.2025.2560191.

[Wi12] E. O. Wilson, ***The Social Conquest of Earth***, Liveright Publishing Cor (2012)

[Yo21] M. Yousefzadeh, C. Henpita, R. Vyas, C. Soto-Palma, Robbins & L. Niedernhofer, "DNA damage - how and why we age?," ***eLife 10***:e62852 (2021); https://doi.org/10.7554/eLife.628

[Yu07] Junying Yu, et al., "Induced Pluripotent Stem Cell Lines Derived from Human Somatic Cells", *Science 318*, 1917 (2007); https://doi.org/10.1126/science.1151526.

4

Tours of the Mitochondrion and mtDNA

This section will describe two imagined guided tours: through a mitochondrion and around its mtDNA ring.

4.1 Tour 1: Into the Mitochondrion

Like Isaac Asimov's ***Fantastic Voyage***, imagine that you have entered a "Shrink Machine" and have had your body height reduced to 1/100,000,000 of a meter. At your new size, a human mitochondrion is a huge object about 100 times larger than you are. You will now have a guided tour of the structure of one of them. There will be eight stops on this guided tour (Fig. 4.1).

4.1.1 Stop 1: The Outer Wall

We will first have a look at the outer membrane wall of the mitochondrion. As we approach it, we see that it is a smooth, *phospholipid* bilayer about 6 nm thick. It is peppered with large *porins* (voltage-dependent anion channels). These porins are protein-lined pores that allow free passage of ions, small molecules, and metabolites with masses of up to about 5 kDa (kilo-daltons). It is rather like a semi-permeable airport terminal with customs officers examining large cargoes. You float through one of these porin tunnels and enter the *intermembrane space,* a narrow space between the smooth outer membrane wall and the curved and wrinkled inner membrane wall.

J. G. Cramer, *How to Live Much Longer*, Copernicus Books, https://doi.org/10.1007/978-3-032-17741-4_4

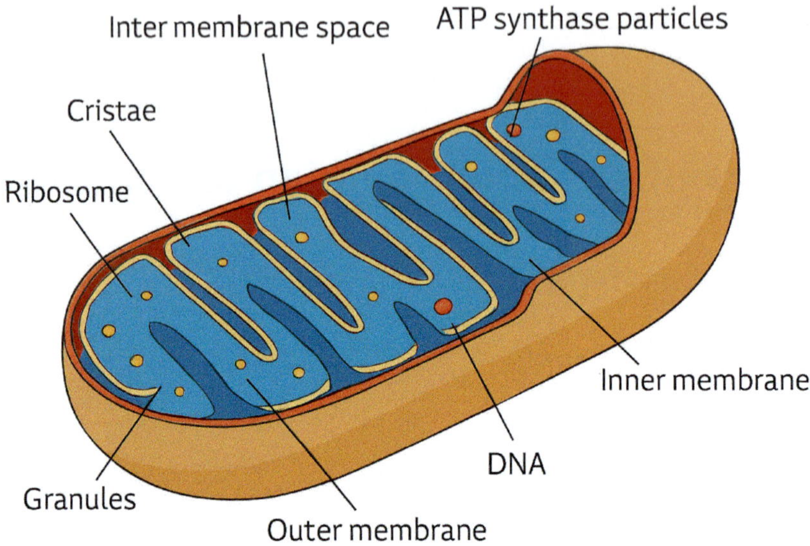

Fig. 4.1 The mitochondrion. *Credit* LadyofHats, Public domain, via Wikimedia Commons

4.1.2 Stop 2: The Inter-membrane Space: The Staging Area

This intermembrane space is teeming with protons (ionized hydrogen atoms) that have been actively pumped into this intermembrane gap to provide basic energy for the cell's respiration process. The proton concentration is much higher here than in the interior region. That creates a steep *proton gradient*. It's a bit like being in a pressurized steam boiler tank or in the gap between the plates of a charged capacitor. Here you also see *cytochrome c* enzymes, compact, reddish globules that can initiate apoptosis (cell death), now waiting in reserve but ready to signal the cell's self-destruct sequence, should it become so badly damaged that it is needs removal and replacement.

Here, proteins dart erratically through the proton soup, occasionally bumping into the inner membrane. You glimpse *Adenylate Kinase* (a smaller enzyme) frantically catalyzing the useful phosphate-swap reaction 2 ADP → ATP + AMP in the crowded space. The sheer density of protons here makes the space feel electrically charged, and it seems to "sizzle".

An acidic tang permeates everything. The viscosity is high, so that movement feels like wading through a thick syrup filled with constantly colliding pebbles (small molecules). The electrical potential difference across the membranes hums ominously. You move downward, approaching the inner membrane.

4.1.3 Stop 3: The Inner Membrane: The Power Grid Wall

You now pass through the inner membrane, which folds itself dramatically into many *cristae,* deep ridges, crevices, and curves like ridges, canyons, and valleys that increase the inner wall surface area. This wall membrane is densely packed with four types of *protein-structured Complexes* that are critical for energy production. Each square micrometer of area holds about 10,000 embedded proteins. It's mostly impermeable, so crossing it without help is like trying to walk through a wall. Therefore, we follow the ATP synthase highway to the next stop.

4.1.4 Stop 4: The Electron Transport Chain: The ATP Assembly Line

We glide along the inner membrane, witnessing the machinery of ATP production at work. Four main protein Complexes are anchored here: *Complex I* (*NADH dehydrogenase*) accepts electrons from NADH (nicotinamide adenine dinucleotide) molecules and pumps protons into the inter-membrane space; *Complex II* (*Succinate dehydrogenase*) receives electrons from $FADH_2$, but doesn't pump protons; *Complex III* (*Cytochrome* bc_1 *complex*) transfers electrons and pumps more protons into the inter-membrane space; and *Complex IV* (*Cytochrome c oxidase*) transfers electrons to oxygen, creating water, and pumps protons into the intermembrane space. These four Complexes consume glucose and oxygen for the energy they use to pump hydrogen ions. They are like the stokers in the boiler room of a great steam ship, busily shoveling in coal to fire the boiler.

Electrons zip along these Complexes like sparks down a power line. Their energy pumps hydrogen ions across the inner membrane, building up the electrochemical gradient. The proton pumping by the Complexes is building up the electric potential between the inner and outer membranes, the sign of a healthy mitochondrion (Fig. 4.2).

4.1.5 Stop 5: ATP Synthase: The Nano Turbine

Now we come to the crown jewel of the mitochondrion. It stands like a multi-story mechanical turbine. It is the *ATP synthase* (*Complex V*) unit, which uses the accumulated proton gradient like pressurized steam to drive its spinning turbine rotor, delivering ATP molecules, powered by the hydrogen ions that move back across the potential gradient, driving the rotor [Mi13]. Every full

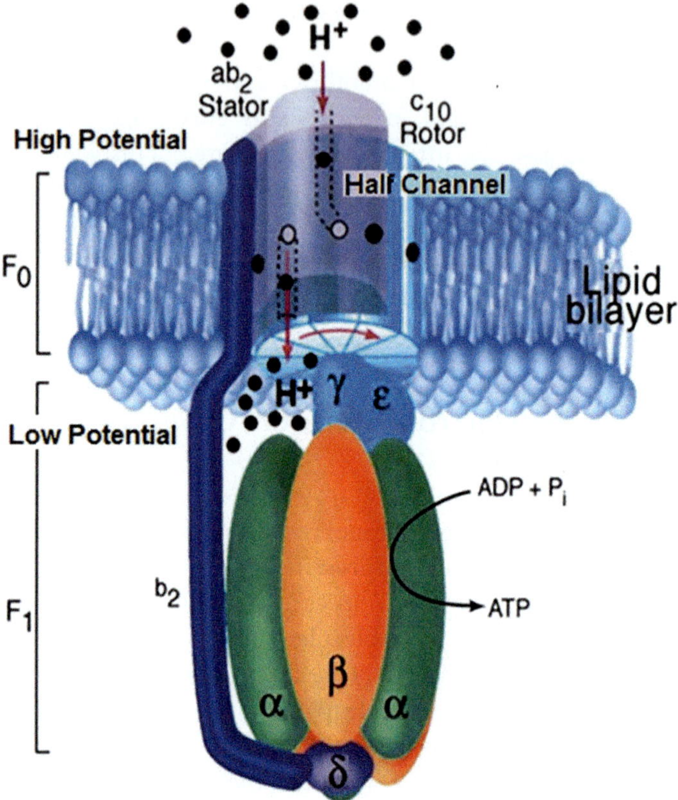

Fig. 4.2 ATP synthase (Complex V). *Credit* Reference [Mi13], used with permission

rotation catalyzes the creation of ATP from ADP and a phosphate molecule. You feel dizzy just watching the rotor spin at about 100 to 200 revolutions per second, delivering one ATP molecule with each revolution. The intermembrane cavity plays the role of a pressurized steam boiler, and this is the rotating turbine it drives. This is where the mitochondrion earns its title: "Powerhouse of the Cell." Typically, in each healthy mitochondrion there are 7,000 or more of these Complex V engines, all busily cranking out ATP molecules as energy carriers to power cellular operations.

4.1.6 Stop 6: The Mitochondrial Matrix: The Chemical Cauldron

You now descend into the *matrix*, the central chamber of the mitochondrion. It is a dense, protein-rich solution filled with *mitochondrial ribosomes* (smaller than cytoplasmic ribosomes), which create the 13 mtDNA-encoded proteins and 24 RNAs. In the deepest cavities of the matrix are *nucleoids* [Bo08] containing several mtDNA loops and their associated transcription and support molecules. In the matrix there are also enzymes of the Krebs Cycle that catalyze the breakdown of *acetyl-CoA,* producing NADH and $FADH_2$ for the electron transport chain to use. It's like a busy chemical kitchen, with bubbling reactions, molecular mixers, and genetic scribes, all operating in concert to support metabolism.

4.1.7 Stop 7: A Mitochondrial Nucleoid: The Genetic Control Room

We now examine one of the *mitochondrial nucleoids* [Bo08, Ka24], which encapsulates cloverleaf clusters of mtDNA bound in loops by TFAM proteins holding them. In each nucleoid there are several copies of mtDNA, usually organized into discrete transcriptional hubs. These are the "mission control centers", engaging in intracellular signaling [Pi22, Pi25a], transcribing messenger RNA, directing the synthesis of mitochondrial proteins, and occasionally replicating an entire mtDNA ring to increase the copy count (Fig. 4.3).

4.1.8 Stop 8: Mitochondrial Dynamics: The Fusion and Fission Zones

Before our tour ends, we hover at the wall junctions where mitochondria split and merge. This dynamic behavior, governed by proteins (mitofusins) like *MFN1/2*, *OPA1*, and *DRP1*, allows mitochondria to join or split as they adapt to cellular energy needs, isolate damaged regions, or share contents. It's like watching a living network of organelles that breathe, divide, and recombine like intelligent machines.

This ends Tour 1 of a mitochondrion. Your guide hopes that you enjoyed the experience and learned from it. Sorry, we cannot accept any tips.

4.2 Tour 2: Around the mtDNA Ring

Now imagine that we use our Shrink Machine once again to reduce your size even further, this time down to the picometer scale. Now you move inside a looming nucleoid for your tour around one of the rings of mitochondrial

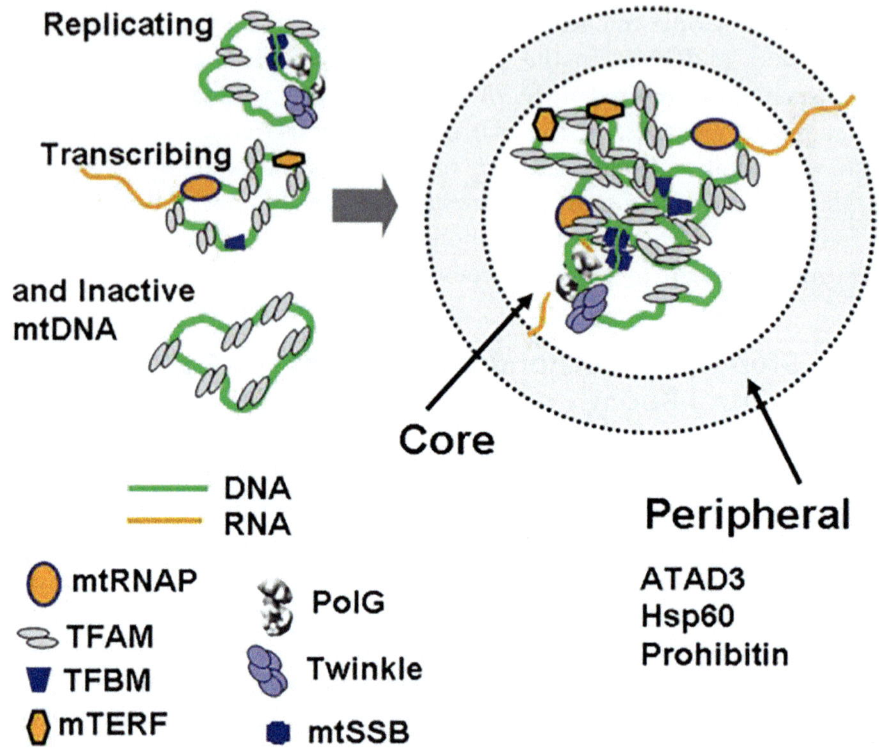

Fig. 4.3 The mitochondrial nucleoid containing mtDNA being read and replicated. *Credit* Reference [Bo08], used with permission

DNA stored there, which has been uncoiled into an open circular ring for your inspection. From your new perspective, the huge double-stranded DNA helix of a human mtDNA (the ring molecule is 16,569 base pairs long) stands before you like a colossal spiraling racetrack.

For position location on the ring, the mtDNA base pairs are numbered in the conventional way, which, for historical reasons, starts with a "1" in the middle of the D-loop. The two intertwined DNA strands are described as the "light" and the "heavy" strand, because of their slightly different total molecular weights. The light strand has a molecular weight of 5063 kDa (kilodaltons) and the heavy strand has a molecular weight of 5174 kDa.

Unlike nuclear DNA, in which all of the protein and RNA coding sequences reside on only one of the two DNA strands, we see that mtDNA shares the coding between the two strands. The light strand contains one protein-coding gene (ND5) and 8 transfer RNA genes. The heavy strand contains the other 12 protein-coding genes, the two ribosomal RNA genes, and 14 of the transfer RNA genes. Note that there are 22 transfer RNAs to transfer just 20 amino acids, because two of those amino acids, serine and leucine, each require two distinct transfer RNAs.

4.2.1 Stop 1: The D-Loop (Displacement Loop): The Command Center

Our tour starts at the *D-loop*, the regulatory hub shown in gray in Fig. 4.4. Curiously, a part of it has an extra DNA strand, making it the only triple-stranded DNA in the human genome. The D-loop spans nucleotides 16,024–16,569 and continues on with 1–576. It is the *control region*, housing the heavy-strand and light-strand promoters (HSP and LSP). Mitochondrial DNA does not have telomeres as landing pads to start DNA replication. Instead, they have *origin points* for the two DNA strands. Here we find the origin point (HO) for the H-strand replication. You'll see *TFAM* (Transcription Factor A, Mitochondrial) binding here, bending the mtDNA into a loop to help initiate transcription. This zone is non-coding, and it's ***highly changeable,*** a "hotbed" for random mutations that are useful for tracking DNA evolution and forensic studies. Here, the *DNA Polymerase γ* and the *mitochondrial RNA polymerase* gather, ready to launch on command to take on the jobs of replication and transcription.

4.2.2 Stop 2: OriL (Light Origin): The Light Strand Launchpad

Now we move clockwise around the ring to about nucleotide 5,770. Here lies *OriL*, the origin point for replication of the light mtDNA strand. It's a short sequence that forms a stem-loop structure when single-stranded, which recruits *DNA Polymerase γ* to initiate the replication. It only activates once the heavy strand has already been replicating long enough to expose OriL. It's a precise timing device, ensuring staggered bidirectional replication.

4.2.3 Stop 3: Protein-Coding Genes: The Power Plant Machinery Blueprints

The next segments are very busy zones encoding the 13 mitochondrial proteins. Let's visit a few key ones:

1. *ND1–ND6, ND4L (NADH dehydrogenase subunits)*: these proteins span various points on the H-strand shown in orange in Fig. 4.4. These are

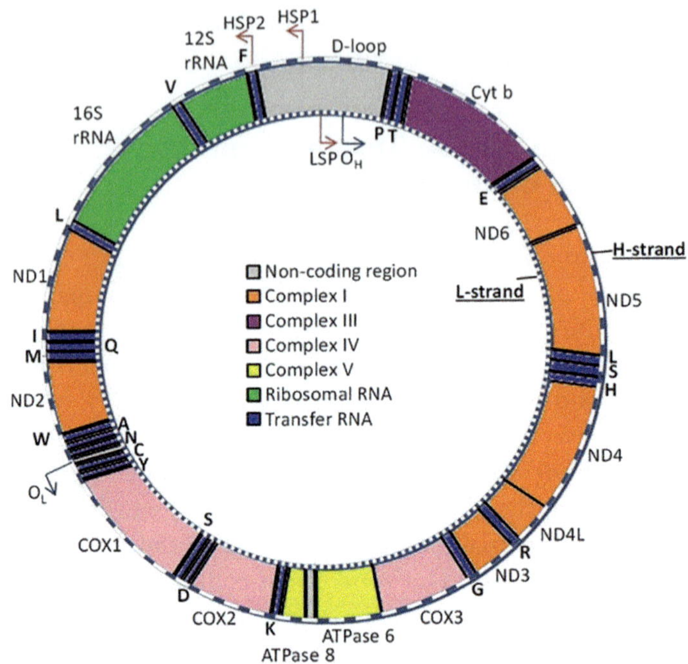

Fig. 4.4 The mitochondrial DNA ring. *Credit* Reference [Wi17], used with permission

blueprints for Complex I of the electron transport chain (see Tour 1), sending mRNA instructions to a ribosome. Inside the matrix, in Tour 1, we had seen ribosomes translating these messages and feeding the constructed proteins into the inner membrane;

2. *CYB (cytochrome b, Complex III)*: around nucleotides 14,747 to 15,887 and shown in purple in Fig. 4.4. A central protein in electron transport. You may also spot *Coenzyme Q10* nearby, shuttling electrons to this station;
3. *COX1, COX2, COX3 (cytochrome c oxidase subunits)*: Blueprints for parts of *Complex IV* are scattered between nucleotides 5,904 and 9,207 and are shown in pink in Fig. 4.4. You'll see translation and immediate membrane insertion, since these proteins are hydrophobic and need to avoid the water in the mitochondrial interior;
4. The genes for *ATP6 and ATP8*, which overlap slightly (8,469–9,207), are shown in yellow in Fig. 4.4. They encode the Complex V *ATP synthase subunits,* essential parts for the spinning turbine of ATP synthase that we saw on Tour 1.

4.2.4 Stop 4: Two rRNA Genes: The Mitochondrial Assembly Line Tools

Here we find *12S rRNA* (648 to1,601) and *16S rRNA* (1,671–3,229), which are shown in green in Fig. 4.4. These are the *toolkits* for building the mitochondrial ribosome, which is smaller and more compact than its main-cell ribosome counterpart in the outside cytoplasm. The two rRNAs are transcribed and combined with imported nuclear-encoded ribosomal proteins to form the full mito-ribosome. If you are familiar with the structure of nuclear DNA, you may notice the lack of introns in the coding here. The mtDNA is more efficient and minimalist, so no introns are present.

4.2.5 Stop 5: The 22 tRNA Genes: The Molecular Transporters

Scattered like pit stops across the mtDNA ring are the 22 *tRNA genes*, each producing a transport protein for a specific amino acid, bringing it to the ribosome or protein assembly. These are shown in blue in Fig. 4.4. The tRNA string folds into a characteristic cloverleaf shape. Some key features in the mtDNA: the tRNA genes flank most of the protein-coding genes, acting like punctuation marks (and sometimes helping to terminate transcription or to mark cleavage sites). These tRNAs are highly specialized to work within the mito-ribosome's unique structure. Some are tightly packed, with barely any space between them, evidence of intense genome compaction. They are the most sensitive part of the mitochondrial genome, because if any of them should suffer even one shape-altering mutation, it is likely that the entire functioning of the mitochondrion would come to a halt. The cause of inherited mitochondrial diseases is often a point mutation that modifies a transport RNA (tRNA) sequence.

4.2.6 Stop 6: Replication Fork and Transcription Bubble: In Action

Along this journey, we encounter:

1. *DNA Polymerase γ*, the only mtDNA polymerase used for replication, proofreading, and building of new mtDNA strands;
2. *TWINKLE helicase*, which has the function of unwinding the mtDNA double helix ahead of transcription;

3. *Mitochondrial single-strand binding proteins (mtSSB)*, which stabilizes loose DNA strands; and
4. *Mitochondrial RNA polymerase (POLRMT)*, which transcribes RNA directly from the DNA template genes. You'll even see *bidirectional transcription*, where both heavy and light strands are read simultaneously in opposite directions from distinct promoters.

4.2.7 Stop 7: Post-transcriptional Processing Sites: The Editor's Desk

Once transcribed, the long precursor RNA is cleaved at tRNA boundaries by enzymes like *RNase P* and *RNase Z*. Each tRNA gets snipped out, while rRNAs and mRNAs are trimmed, matured, and polyadenylated (adding a string of adenine nucleotides). It's like watching a film editor cut and splice a reel, preparing the RNA for export and translation.

4.2.8 Stop 8: Mutation Hotspots and Damage Control Units

Due to transcription errors and the proximity to the electron transport chain, which produces ATP and its ROS byproduct, some mtDNA regions, especially in the D-loop and protein-coding genes, show frequent mutations. You will see base excision repair enzymes like *OGG1* cleaning up some of the oxidative damage. However, mtDNA lacks robust mismatch repair, so many errors persist, contributing to aging and disease.

As you complete your 16,569 base-pair circumnavigation of the mtDNA, you will have just traveled past the only genetic material that human mitochondria retain from their ancient bacterial ancestry. The rest of their extensive bio-machinery has been outsourced and is now encoded by the nuclear genome.

This surviving closed loop of DNA is a vestige of previous bacterial independence, but it is also the site of ongoing evolution, interaction, regulation, mutation, and metabolic history. As long as the loop remains intact and unmodified, the mitochondrion it supports stays healthy and productive, the host cell has plentiful energy, and the organism is healthy. When the loop has sections removed by deletion mutations or suffers random changes from point mutations, its support of the mitochondrion is compromised, an ATP energy shortage sets in, and this can lead to the collapse of the whole system. We call that process "human aging".

References

[Bo08] D. F. Bogenhagen, D. Rousseau, and S. Burke, "The Layered Structure of Human Mitochondrial DNA Nucleoids," ***Journal Of Biological Chemistry 283*** (6), 3665–3675 (2008); https://doi.org/10.1074/jbc.M708444200.

[Ka24] E. V. Kakudji and S. C. Lewis, "Mitochondrial nucleoids," ***Current Biology 34*** (21): R1067–68 (2024).

[Mi13] J.H. Miller, Jr, K. I. Rajapakshe, H. L. Infante, and J. R. Claycomb, "Electric Field Driven Torque in ATP Synthase,". ***PLoS ONE 8(9***): e74978 (2013). https://doi.org/10.1371/journal.pone.0074978.

[Pi22] Martin Picard, et al, "Mitochondrial signal transduction," ***Cell Metabolism***, (2022).

[Pi25a] Martin Picard, "Mitochondria Are More Than Powerhouses - They're the Motherboard of the Cell," ***Scientific American***, May 20, 2025.

5

Hallmarks of Aging, Aging Clocks, and Aging Tests

5.1 The Nine (or Twelve or Twenty One) Hallmarks of Aging

In 2013, López-Otín, Serrano, Partridge, Blasco, and Kroemer published a seminal paper defining the *Nine Hallmarks of Aging,* as illustrated in Fig. 5.1. [Lo13]. They divided their Hallmarks into three functional classes: (A) damage causes, (B) damage responses, and (C) culminating effects. In the *causes* category were (1) genomic instability (increasing mutations), (2) telomere attrition (shortening), (3) epigenetic alterations (age-based switching on and off of genes) [Bo23], and (4) loss of proteostasis (protein balance). In the *responses* category were (5) deregulated nutrient sensing, (6) mitochondrial dysfunction, and (7) cellular senescence (damaged cells that cease dividing and produce inflammation). In the *effects* category were (8) stem cell exhaustion and (9) altered intercellular communication. Essentially, the authors were "collecting stamps" as represented by these nine ways in which cells are observed to fail, since they did not suggest cause-and-effect relationships between any of their Hallmarks. Further, in my opinion, they have the basic hierarchy completely wrong, because *mitochondrial dysfunction* and the cellular energy shortage it creates are hardly a *response* to damage at all, but rather are the hidden driver and root cause behind all of the other Hallmarks [We25c]. In Fig. 5.1, I have rearranged the nine Hallmarks to put what I consider to be the most fundamental ones at or near the top.

In 2023, the same authors did another Hallmark round, proposing an expansion of the original Hallmarks list to include 12 items, which would now be expanded to include: (10) disabled macroautophagy (failed removal

J. G. Cramer, *How to Live Much Longer*, Copernicus Books, https://doi.org/10.1007/978-3-032-17741-4_5

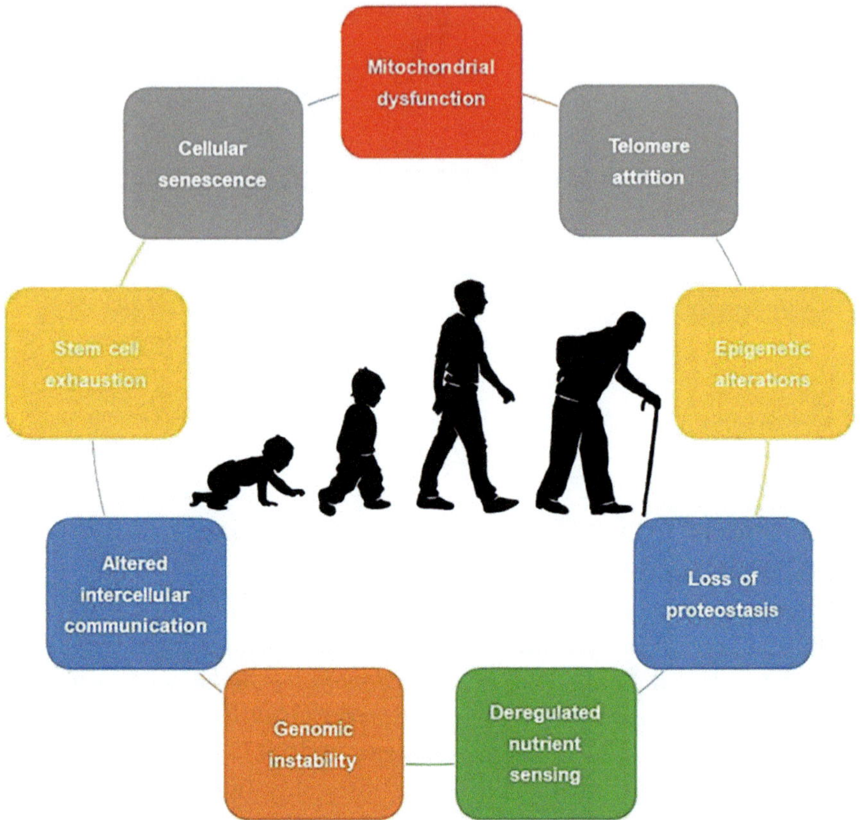

Fig. 5.1 The original Nine Hallmarks of Aging. *Credit* Wikimedia Commons, modified by JGC

of damaged cells), (11) chronic inflammation, and (12) compromised cell identity (cells forgetting their function). Again, these are presented without suggesting any causal connection to the other Hallmarks. More "stamps" for the collection.

Later authors, perhaps because their own bio-specialties were not included in the original Hallmarks list, have argued strongly for inclusion of even more Hallmarks. These include: (13) microbiome symbiosis (disruption of gut bacteria), (14) iron dyshomeostasis (disruption of normal iron levels) [Bi20, Ro20], (15) lipidome imbalance, (16) NAD+ depletion, (17) splicing dysregulation, (18) altered mechano-transduction, (19) immuno-senescence (failure of immune cells), (20) mitochondrial heteroplasmy (growing mtDNA fraction with mutations), and (21) clonal expansion of mtDNA mutants (exponential growth of short mtDNA rings containing deletions).

I prefer the original list of nine Hallmarks, while expanding it to 21 items or so seems rather excessive, particularly because there is no distinction made between causes and effects. In my opinion, essentially all of these Hallmark "stamps" are downstream effects of the devastating shortage of cellular ATP energy that comes from exponentially rising mtDNA damage with increasing age. Initially, evolution has supplied elaborate cellular maintenance and repair mechanisms for cells and for nuclear DNA and damage removal by autophagy to correct for most of the maladies identified by the Hallmarks. However, if the ATP energy produced by the available mitochondria cannot support these mechanisms, they go offline and remain dormant. Then the mitochondrial damage builds, and the other ugly Hallmarks appear.

5.2 How Old Are your Body Parts?

One critical element needed in considering the issues of longevity and aging is to have available quantitative ways of measuring the varying biological ages of body parts: blood, skin, intestines, liver, kidneys, pancreas, bone marrow, heart, brain, ... In evaluating the effectiveness of longevity interventions that can slow, halt, or reverse aging, we would like to focus on how we can *know* if a particular intervention is actually doing anything beneficial by observing before-and-after test results. Unfortunately, in the present health area, personal anecdotal "feel-good" endorsements are much more common than actual quantitative test results. Effectively, we are flying blind.

Many of the longevity interventions discussed here, if tested at all, have been tested mainly using animal models, usually laboratory mice. In particular, the standard laboratory mouse is widely used in age-related experiments. These have an average lifetime of about 26-30 months and a maximum lifetime of about 3 years. The age-related effects that set the hard limit on the maximum lifetime in mice are not known but may be related to tumor development. However, that maximum lifetime for short-lived mice (~ 3 years) is likely to have a quite different set of root causes from the maximum lifetime for long-lived humans (~ 120 years).

Nevertheless, the researchers investigating interventions that may combat the effects of human aging remain very focused on the impact of their interventions on the lifetime, health-span, and maximum lifetime of their laboratory mice. This, of course, makes possible relatively short test-to-publication cycles. It is also fairly standard to test and record the state and changes in observed mental acuity and physical performance of the animals as interventions are being tested.

In addition, many tests on humans take the easy route in obtaining samples. Because it may be difficult and complicated to obtain representative tissue samples from most body organs, the tests almost universally focus on the easy-to-obtain samples of blood, urine, saliva, and nasal and cheek swabs. For example, the determination of telomere length and senescent cell count is almost always done with white blood cells only. Saliva provides neutrophil white blood cells and a few epithelial (skin) cells. Urine samples contain some detached kidney cells. There is little research that supplies any information about how closely these tests using easy-to-obtain tissue samples accurately reflect the condition of the more inaccessible body parts and organs.

5.3 Fitness and Cognition Testing

Researchers are also investigating specific measures of fitness and cognition as potential biomarkers of biological age. Physical fitness is closely linked to aging because mobility, strength, endurance, and lung function tend to decline over time. Cognitive function is another key marker of biological age. Declines in memory, processing speed, pattern recognition, and executive function can indicate accelerated aging. These markers can indicate the physiological state of an individual's aging process, rather than just how many years they have lived.

5.3.1 Cardiorespiratory Fitness (CRF)

Often measured by VO_2 max (maximal oxygen uptake, see Section 5.8 below), CRF is a strong predictor of cognitive health. Higher CRF is consistently associated with better performance across multiple physical and cognitive domains, particularly those susceptible to age-related decline (e.g., executive function, memory, processing speed). Some studies even show a relationship between high CRF and an increased in volume of the brain's hippocampus.

5.3.2 Muscular Strength

Measures like grip strength and lower-body power are also linked to cognitive function, though the relationship might be less direct than with aerobic fitness. Maintaining muscle strength is crucial for overall functional independence in older adults.

5.3.3 Balance and Gait

These physical measures reflect the neurological integrity and condition of the inner ear vestibular system, which are often impacted by aging. Declines in balance and gait and the onset of vertigo when first standing up are associated with cognitive decline and increased fall risk.

5.3.4 Cognitive Test Batteries

Comprehensive neuro-psychological test batteries that assess various cognitive domains (e.g., working memory, attention, logical reasoning, processing speed, verbal memory) can provide a detailed picture of an individual's cognitive status. The rate of decline in these measures can be a significant indicator of accelerated or healthy aging.

5.4 Bloodwork and Phenotypic Age

Researchers are identifying blood-based biomarkers that are influenced by exercise and that correlate well with cognitive function. Examples include *Cathepsin B* (CTSB), a lysosomal enzyme secreted from muscle after exercise, associated with memory function; *Klotho*, a circulating protein linked to enhanced cognition and synaptic function; and *Inflammatory Markers*, particularly the levels of pro-inflammatory markers *IL-6*, *TNF-α,* and *C-reactive protein.*

Another common blood-related measure is to observe before-and-after values of blood-based quantities that are readily measurable in commercial tests from a blood sample, e.g., the concentrations of albumin, creatinine, glucose, alkaline phosphatase, and C-reactive protein, as well as lymphocyte count, white blood cell count, mean cell volume, and red cell size distribution width. A paper [Le18] by Levine et al., presented a procedure for combining these blood-work values to obtain the subject's "phenotypic age", i.e., the apparent biological age as implied by blood variables, and a spreadsheet is available online for performing the calculations. Results show that this easily accessible quantity correlates well with other measures of biological age.

5.5 Noise as a Clock: Protein Expression Variance

In their work on plasma dilution as a longevity intervention, the Conboy Group at U. C. Berkeley [Ki22] identified ten marker proteins, the variance or "noise" in the expression of which can be used as a measure of biological age. In particular, they found that in producing the distribution functions of the degree of expression of these ten proteins from cell to cell, the standard deviations characterizing the widths of the distribution curves correlated very well with other measures of biological age. Apparently, the random noise in protein production grows proportionately with age, implying that the epigenetic silencing of certain protein-coding genes with increasing age has a growing stochastic aspect.

5.6 Telomere Length Assay

As will be discussed further in Chap. 8, Sect. 8.3, telomeres are special nuclear DNA sequences at the ends of chromosomes that provide a "docking zone" for the enzymes that perform nuclear DNA replication during cell division. Telomeres grow shorter with each cell division (because the enzyme docking region itself is not replicated) until they may become too short to function. When that happens, cell division is blocked, chromosomes may fuse, and the cell becomes senescent or dies.

There are now standard tests that measure telomere length in white blood cells taken from a blood or saliva sample. These tests are intended to assess the overall telomeric condition of the subject. For what it is worth, white-blood-cell telomere length can be viewed as a rough indication of biological age, but it has *not* been found to correlate well with most of the other bio-clock age measurements.

At present, it is not clear if the telomere length for white blood cells [Ca14] can provide an accurate representation of the telomere length condition for all of the other cell types, particularly those with high replication rates (e.g., the digestive system, where cell damage is frequent and turnover is high).

Also, new white blood cells are continually being produced by the stem cells in bone marrow and undergo constant quality control, so that only the best are likely to appear in a sample. These may not be very representative of slow-dividing long-lived cells (e.g., liver and kidney cells) elsewhere in the body. Blood cells may be among the easiest cell types to sample, but other cell types (e.g., from muscle cell micro-biopsies or the loose kidney cells in a urine sample) may be more representative.

5.7 Epigenetic Programming and CpG Methylation

Epigenetic age assays have lately become a widely used standard representation of biological age. To understand what that is, let's begin with a rough description of the mechanism behind epigenetic programming in cells. The "human genome" means the net DNA content, present in all cells, which is encoded using the three-letter GCAT sequences of nucleotides. These sequences usually encode instructions for assembling all needed proteins, the folded chains of amino acids that are the fundamental structures of life. These assembly instructions are stored in the DNA of the 23 chromosomes inherited from the father and the 23 chromosomes from the mother, containing over three billion nucleotide base pairs, forming about 49,000 genes. But many of these genes are "switched off".

Most of the protein-coding genes within the nuclear DNA are preceded by a promoter region and a 5′ untranslated region (5′UTR in Fig. 5.2), essentially functioning as an ID code used by the cell's transcription enzymes to find and identify the gene and its DNA code to be transcribed for protein production. One can think of this as if each gene is a book stored in the DNA library, with the title and call number on the book's spine represented by the gene's promoter region. Some of these books are locked away on inaccessible shelves (histone spooled DNA), while other books are stacked or open and ready for use on the library table.

Typically, each promoter region of a gene contains a cluster of repeating sequences of the nucleobases cytosine (C) followed by guanine (G) (see Appendix A). These are called "CpG islands", with "p" indicating the phosphodiesterase bond (see Fig. B.2 in Appendix B below) in the supporting DNA backbone. The significant feature of the CpG sequences is that the cytosine base at its fifth carbon position can be methylated [Ch21], i.e., a carbon-hydrogen CH_3 methyl group can be attached to it in place of a hydrogen atom.

Interestingly, in mitochondria the expression of the genes in the mtDNA does not seem to be controlled in the same way. Researchers [Wi17] have found that when CpG methylation occurs in mtDNA, it has no effect. Curiously, "GpC" methylation in mtDNA does produce mild changes in mitochondrial gene expression. However, the 13 protein-coding genes of

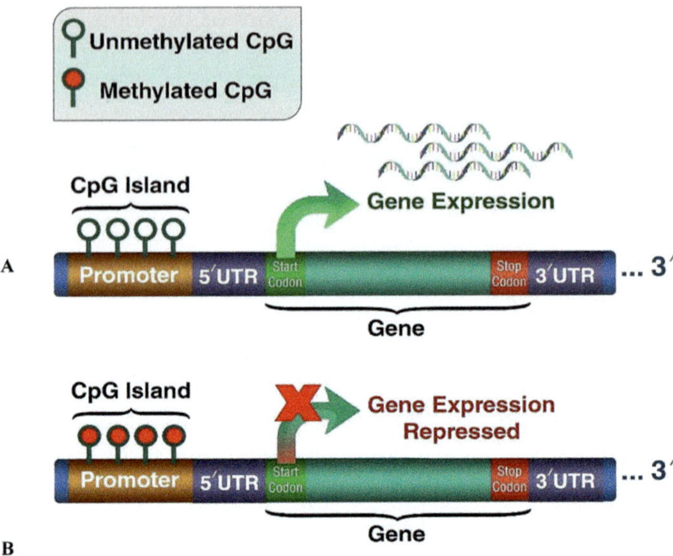

Fig. 5.2 Methylation of CpG sites switches off gene expression. *Credit* Reference [Ch21], used with permission

mtDNA are normally "always on" because of their important roles in providing components for cellular ATP energy production.

When CpG methylation is present in nuclear DNA, it encourages that region of the DNA chain to "close" and to be wrapped around the spool-like histone structures for compact inactive storage, and it also discourages any wandering transcription factors from landing on the promoter region and transcribing the DNA coding. When a sufficient number of methylations are present in the CpG islands of a promoter, this has the effect of completely silencing the gene, rolling the DNA sequence up on a histone, and preventing production of its encoded protein. In other words, while all body cells have the same genes, ready to express their proteins, only those genes with the promoter region relatively clear of methylation will actually express their proteins, and the rest are silenced. This selective methylation is the principal epigenetic mechanism by which cells become specialized to perform specific functions and also by which cells are "set" to be either young or old.

DNA methylation measurements now provide very important and widely used analytic markers, because bio-equipment suppliers like Illumina have developed CpG methylation microarrays that allow simultaneous determination of the methylation condition of many thousands of CpG sites within a DNA sample. In 2013, UCLA Prof. Steve Horvath used data from two sizes of Illumina microarrays that identify the methylation states of 27,000 CpG sites [Ho13]. Using advanced statistical techniques and machine learning, he isolated 353 CpG sites that were highly correlated with human calendar age. This analysis was based on 7,844 tissue samples that spanned 51 different tissue types. He found that CpG methylation *decreased* with age in 55% of these sites and *increased* with age in the other 45%. In other words, as the human subjects aged, some genes were being systematically silenced and others activated. Some genes were being brought into operation for protein production, and others were being taken offline. This work became known as the "Horvath DNAm Clock" and has become a standard for aging research. In 2018, Horvath improved the technique with a new clock algorithm [Ho18] using 513 CpG sites from analysis based on even larger Illumina microarrays (450 k and 850 k).

One central question raised by the discovery of this methylation clock is: *Is this epigenetic programming the* ***result*** *of aging, or is it the* ***cause*** *of aging?* In other words, if one could provide a subject with a treatment that would reset the Horvath Clock to a younger nuclear DNA methylation profile, would that action rejuvenate the subject? (Or would it be the equivalent of attempting to time-travel into the past by setting back your kitchen clock, or expecting your car to run like new if you reset its odometer to zero?)

The surprising answer, based on several animal experiments, seems to be that an epigenetic reset of the methylation pattern *does* produce some degree of actual rejuvenation. Further, since a methylation profile tends to remain relatively unchanged once it is established, any intervention that produces a DNA methylation change or reset to a younger profile can be taken to be relatively permanent and will probably last for a period measured in years or decades.

One interesting question that this research raises is this: *Why does the human body's epigenetic programming profile change with increasing age?* What is the overall system responding to in modifying our epigenetic profile as we age? There is some evidence that many of the genes that are silenced by age-related epigenetics are those coding for proteins that are associated with large ATP energy consumption. The epigenetic mechanisms of the body appear to protect the brain [Br25] and heart, both heavy users of ATP energy, at the expense of many other organs.

In my opinion, this suggests that what is going on with the age-dependent shift in epigenetic programming is, at least in part, a response to the growing shortage of ATP energy [Na08, Pi14, Sm08]. As the mtDNA is accumulating damage, the body is silencing genes associated with large energy consumption and bringing online some less effective substitute genes that consume less energy. This is perhaps an oversimplification, in that some of the epigenetic age shift may also be prompted by other factors like responses to stress, disease, and changing conditions of environment and diet. Nevertheless, I consider the energy deficit to be the main driver of human aging.

5.8 Energy Use: Oxygen Consumption, VO_2 Rest and Max

One way of determining whether the cellular mitochondria overall are performing properly is to measure the rate at which atmospheric oxygen is being consumed, because the body's mitochondria are the primary consumers of that oxygen. Manufacturers like Calibre Biometrics [Ca25a] now produce a relatively inexpensive system, consisting of a biometric mask with a linked cellphone app, that can deliver consistent measurements of the amount of oxygen that the subject consumes in a given situation. Such measurements are usually a VO_2 *rest* measurement, in which the baseline O_2 consumption of the subject is measured in a resting situation, and a VO_2 *max* measurement, in which the subject uses a treadmill to perform vigorous exercise until maximum O_2 consumption of the subject is reached. The question is then, what do these observations of oxygen consumption actually tell us about mitochondrial condition? VO_2 max measures not only the ability of cells to use oxygen, but also the ability of the lungs, heart, and bloodstream to deliver the oxygen to muscle cells and elsewhere for the O_2 consumption. On the other hand, VO_2 rest is perhaps a meaningful measure of the average condition of mitochondria that are not under exercise-induced stress and perhaps provides more meaningful information about the intrinsic mitochondrial condition. We can expect to learn more about these issues as new data is collected, and several different measures of mitochondrial condition can be compared.

5.9 Senescent Cell Testing

As discussed above, the seventh of the original nine Hallmarks of Aging is *Cellular senescence*, the condition in which a damaged cell is "retired in place" and prevented from further cell division, with the consequences that it causes inflammation and emits toxic SASP (Senescence-Associated Secretory Phenotype) that affects its cellular neighbors. In doing senolytic sessions to clear such senescent cells, as described in Chap. 6, it would be very useful to be able to perform before-and-after measurements of the body's average senescent cell fraction, as an indication of whether the senolytic intervention is actually removing senescent cells as expected (Fig. 5.3).

Unfortunately, determining the presence and fraction of senescent cells in a test subject is not easy. The most widely used test for the presence of cellular senescence is done by taking a tissue biopsy, preparing it as a microscope slide, and staining that to highlight cells expressing the protein *SA-β-gal* (Senescence-Associated β-galactosidase), a well-established senescent cell marker. The cells that are senescent will then appear blue when viewed in the microscope, so that they can be counted to estimate the fraction of senescent cells in the sample. This procedure, while useful in a lab setting, is not appropriate for general testing to provide individuals with a quick estimate of their senescent cell fraction.

There is, however, another technique that would be useful but is not yet generally available. Senescent cells are known to be prolific sources of SASP, toxic secretions that should appear in the bloodstreams of individuals with a significant burden of senescent cells. A senescent cell test now being investigated would be to perform a complete SASP Biomarker Panel on a blood sample, if such were available. This would involve measuring the concentrations of the many senescence biomarkers, including the SASP component pro-inflammatory cytokines, chemokines, and some growth factors, and

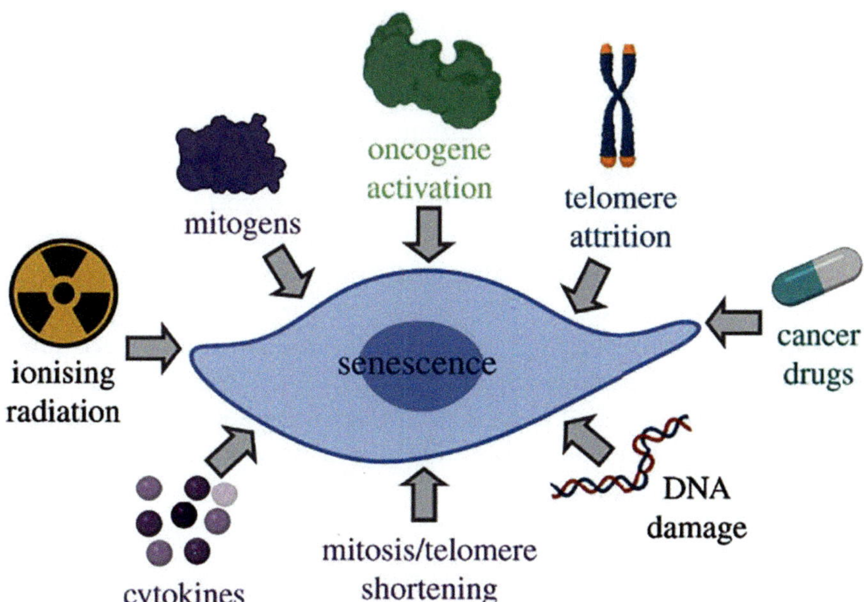

Fig. 5.3 Senescent cells and their causes. *Credit* GNU Free Document License

establishing calibrations to interpret these results. Although testing companies like LabCorp [La25] offer individual tests for most of the significant SASP markers, no comprehensive SASP Biomarker Panel obtained from a blood sample is currently commercially available. Vendors take note!

5.10 The mtDNA Copy Number and ddPCR Testing

"Mitochondrial DNA copy number" refers to the number of mtDNA copies that are present in an average cell. This quantity varies depending on tissue type, metabolic demand, and physiological conditions. It serves as a biomarker for mitochondrial function (the more the better), and its decrease has been linked to aging, disease progression, and cellular health.

The PCR (Polymerase Chain Reaction) was developed in 1983 by Kary Mullis. It is an amazingly powerful technique that has made possible many advances in molecular biology. The widely used laboratory technique amplifies DNA and RNA by heat cycling them in the presence of special primers and polymerases, doubling their number with each heat cycle.

Droplet digital PCR (ddPCR) is a highly sensitive variant method for quantifying mtDNA copy number. Unlike traditional quantitative PCR, ddPCR partitions the DNA sample and replication enzymes into many thousands of tiny nanoliter-size droplets, uses heating cycles to implement PCR amplification in all of these, and then digitizes the number and fraction of the droplets that are now fluorescent, indicating that they contain mtDNA after the amplification. This new technique allows for absolute quantification of mtDNA copy number without reliance on standard curves or extrapolations. The technique enables precise measurement of mtDNA copy number down to single-cell resolution, making it valuable for studying the presence of multiple mtDNA variants within a cell and for accurately estimating the degree of mtDNA damage.

Researchers have used ddPCR to assess mtDNA copy number variations in contexts such as cancer progression, neurodegenerative diseases, and immune cell activation. The method is particularly useful for detecting subtle differences in mtDNA abundance that might be indistinguishable using more conventional techniques.

5.11 MitoClock mtDNA Damage Assay: Deletions and Point Mutations

Mitrix Bio [Mi25] has developed a "MitoClock" test procedure for determining the condition of the mtDNA in a sample of mitochondria. The technique is sensitive to both point and deletion mutations of the mtDNA and shows them separately. Using the newly available long-range PCR and nanopore sequencing technologies, they employ a proprietary machine learning technique to quantitatively characterize the degree of DNA damage present in the mitochondria from a subject's urine, saliva, or blood samples. In one example, they have shown damage plots for the same mtDNA pattern taken from urine samples of three maternal-line subjects that at the time of testing were 32, 62, and 93 years old. The analysis indicates mtDNA damage scores of 7%, 11%, and 25%, respectively.

There are indications that a subject is not likely to survive long if the general mitochondrial damage level rises above about 25%, because energy production by the mitochondria becomes too inefficient to sustain life. In other words, there is now direct evidence that mitochondrial damage, which will severely limit the energy available to otherwise healthy cells, increases with age and may be the principal contributor to the maladies of old age. Further, there is now a new "clock" technique that can measure this directy. Using this technique, I have arranged for MitoClock measurements of my own mtDNA damage in samples of saliva and urine that I provided. The saliva measurement primarily reflects the condition of neutrophil white blood cells, while the urine measurement primarily reflects the condition of the mitochondria from free kidney cells.

The results showed almost no major mitochondrial damage in my saliva sample, while considerable damage was shown in my urine sample, which was intermediate in damage between the 62 and 93-year-old subjects mentioned above. In particular, the urine sample had somewhat more point mutations and considerably more deletion mutations than the saliva sample. There is a lesson to be learned here. For the same 90-year-old subject (me), the mtDNA of the kidney cells has sustained far more damage than has the mtDNA of the white blood cells. *How can that be?*

The answer seems to be that kidney cells are slower-dividing, have been functioning for longer, and are functioning in an active environment that has very high energy demands, probably resulting in greater production of ROS. Further, kidneys filter waste and toxins that can damage nearby mtDNA. White blood cells in saliva are newly made, tightly controlled for quality, and come from fast dividing cells in an environment with relatively low metabolic activity. Clearly, the degree of damage to mtDNA is very cell-type dependent. Table 5.1summarizes what we have discussed in this chapter.

Table 5.1 Key biological age biomarkers and tests

Biomarker category	Specific test/method	Description and purpose	Sample type	Key insight/ limitation
Fitness and cognition	Cardio-respiratory fitness (VO_2 max)	Measures maximal oxygen uptake; a strong predictor of cognitive and overall health	Physical test	Reflects health of heart, lungs, and circulatory system; linked to brain health
	Muscular strength (e.g., grip strength)	Assesses muscle power; linked to functional independence and cognitive function	Physical test	Relationship to aging may be less direct than with aerobic fitness
	Balance and gait analysis	Evaluates neurological integrity and vestibular system function	Physical test	Declines are associated with cognitive decline and increased fall risk
	Cognitive test batteries	Comprehensive assessments of memory, processing speed, reasoning, etc.	Cognitive test	Rate of decline is a significant indicator of accelerated or healthy aging
Blood-based biomarkers	Phenotypic age (Levine et al.)	Algorithm combining blood values (e.g., albumin, creatinine, CRP) to estimate biological age	Blood	Easily accessible and correlates well with other measures of biological age
	Inflammatory markers (IL-6, TNF-α, CRP)	Measures levels of pro-inflammatory cytokines and proteins	Blood	Chronic inflammation ("inflammaging") is a key hallmark of aging
	Specific proteins (e.g., Cathepsin B, klotho)	Levels of certain exercise-influenced proteins linked to cognitive function	Blood	Potential biomarkers, but not yet standard in commercial panels
Cellular and molecular	Epigenetic clock (Horvath clock)	Measures methylation status at specific CpG sites in DNA to estimate biological age	Blood, tissue	A gold standard. Debate continues on whether it is a cause or result of aging

(continued)

Table 5.1 (continued)

Biomarker category	Specific test/method	Description and purpose	Sample type	Key insight/ limitation
	Telomere length assay	Measures the length of telomeres in chromosomes.	White blood cells (blood)	A rough indicator of biological age, but does not correlate well with other clocks. May not be representative of all tissues
	Senescent cell load (SA-β-gal stain)	Stains for senescence-associated β-galactosidase to identify senescent cells	Tissue biopsy	The gold standard method, but invasive and not practical for routine testing
	SASP biomarker panel	Measures the senescence-associated secretory Phenotype (e.g., cytokines, chemokines) in blood	Blood	Proposed non-invasive test, but a comprehensive commercial panel is not yet available
	Protein expression variance	Measures the "noise" (standard deviation) in the expression levels of specific proteins	Blood (plasma)	Increased noise with age suggests stochastic epigenetic silencing
Mito-chondrial function	mtDNA copy Number (ddPCR)	Uses droplet digital PCR to absolutely quantify the number of mtDNA genomes per cell	Blood, tissue	Biomarker for mitochondrial function; linked to aging and disease
	mtDNA damage assay (MitoClock)	Uses long-range PCR and nanopore sequencing to quantify point and deletion mutations in mtDNA	Urine, saliva, blood	Directly measures mitochondrial damage, a proposed root cause of aging. Shows variation between tissue types
	Oxygen consumption (VO_2 rest and max)	Measures baseline and maximum rate of oxygen consumption	Physical test (mask)	VO_2 max reflects systemic oxygen delivery; VO_2 rest may better reflect intrinsic mitochondrial health

References

[Bi20] G. E. Billman, "Homeostasis: The Underappreciated and Far Too Often Ignored Central Organizing Principle of Physiology," ***Frontiers in Physiology 11***:200 (2020); https://doi.org/10.3389/fphys.2020.00200N.

[Br25] Nora Bradford, "First map of human brain mitochondria is 'groundbreaking' achievement. Hundreds of cubes of human brain tissue help scientists to chart the energy-making capabilities of various brain regions," ***Nature***, News, 26 March 2025.

[Ca14] D. Campa, et al, "Leukocyte telomere length in relation to pancreatic cancer risk: a prospective study," ***Cancer Epidemiol Biomarkers Prev 23***(11): 2447–2454 (2014).

[Ca25a] Calibre Biometrics Inc., 35 Walnut Street, Suite 100, Wellesley Hills, MA 02481, USA; https://calibrebio.com/.

[Ch21] N. Chenarani, A. Emamjomeh, A. Allahverd, S. Mirmostafa, M. H. Afsharinia, and J. Zahiri, "Bioinformatic tools for DNA methylation and histone modification: A survey," Genomics 113 (3), 1098–1113, (2021); https://doi.org/10.1016/j.ygeno.2021.03.004.

[Ho13] Steve Horvath, "DNA methylation age of human tissues and cell types", ***Genome Biology 14*** (10), R115 (2013); https://doi.org/10.1186/gb-2013-14-10-r115.

[Ho18] S. Horvath, J. Oshima, G. M. Martin, A. T. Lu, A. Quach, H. Cohen, S. Felton, M. Matsuyama, D. Lowe, S. Kabacik, J. G. Wilson, A. P. Reiner, A. Maierhofer, J. Flunkert, A. Aviv, L. Hou, A. A. Baccarelli, Y. Li, J. D. Stewart, E. A. Whitsel, L. Ferrucci, S. Matsuyama, and K. Raj, "Epigenetic Clock for Skin and Blood Cells Applied to Hutchinson Gilford Progeria Syndrome and *Ex Vivo* Studies," ***Aging (Albany NY) 10*** (7): 1758 (2018).

[Ki22] D. Kim, D. D. Kiprov, C. Luellen, M. Lieb, C. Liu, E. Watanabe, X. Mei, K. Cassaleto, J. Kramer, M. J. Conboy, and I. M. Conboy, "Old plasma dilution reduces human biological age: a clinical study," ***Geroscience 44*** (6), 2701-2720 (2022); https://doi.org/10.1007/s11357-022-00645-w.

[La25] LabCorp; online https://www.labcorp.com/.

[Le18] M. E. Levine, A. T. Lu, A. Quach, B. H. Chen, T. L. Assimes, S. Bandinelli, L. Hou, A. A. Baccarelli, J. D. Stewart, Y. Li, E. A. Whitsel, J. G. Wilson, A. Reiner, A. Aviv, K. Lohman, Y. Liu, L. Ferrucci, and S. Horvath, "An epigenetic biomarker of aging for lifespan and healthspan," Aging (Albany NY) 10 (4), 573–591 (2018); https://doi.org/10.18632/aging.101414.

[Lo13] C. López-Otín, M. A. Blasco, L. Partridge, M. Serrano, and G. Kroemer, "The Hallmarks of Aging," ***Cell* 153 (6)**, 1194–1217 (2013).

[Mi25] Mitrix Bio Inc., 4695 Chabot Dr. Suite 200, Pleasanton, CA 94588, https://mitrix.bio/.

[Na08] Robert K. Naviaux,:"Mitochondrial control of epigenetics," ***Cancer Biology & Therapy, 7:8***, 1191–1193 (2008); https://doi.org/10.4161/cbt.7.8.6741.

[Pi14] M. Picard, J. Zhang, S. Hancock, O. Derbeneva, R. Golhar, P. Golik, S. O'Hearn, S. Levy, P. Potluri, M. Lvova, A. Davila, C. S. Lin, J. C. Perinh, E. F. Rappaport, H. Hakonarson, I. A. Trounce, V. Procaccio, and D. C. Wallace, "Progressive increase in mtDNA 3243A>G heteroplasmy causes abrupt transcriptional reprogramming, ***PNAS online*** (2014); www.pnas.org/cgi/doi/10.1073/pnas.1414028111.

[Ro20] J. Rosen, et al, "Reverse Transcriptase—Impact on redox homeostasis," ***Redox Biology*** (2020).

[Sm08] D. Smiraglia, M. Kulawiec, G. L. Bistulfi, S. Ghoshal, and K. K. Singh. "A novel role for mitochondria in regulating epigenetic modifications in the nucleus," ***Cancer Biology & Therapy, 7:8***, 1182–1190 (2008), https://doi.org/10.4161/cbt.7.8.6215.

[We25c] V. Weissig, "Mitochondrial dysfunction as the "mother" of all hallmarks of aging," ***Journal of Mitochondria, Plastids and Endosymbiosis, 3:1***, 2560191 (2025); https://doi.org/10.1080/28347056.2025.2560191.

[Wi17] M. G. van der Wijst, A. Y. van Tilburg, M. H. J. Ruiters, and M. G. Rots, "Experimental mitochondria_x005F_xfffe_targeted DNA methylation identifies GpC methylation, not CpG methylation, as potential regulator of mitochondrial gene expression," ***Nature Scientific Reports*** *7*: 177; https://doi.org/10.1038/s41598-017-00263-z.

6

The Longevity Handymen: Short-Term Fixes and Patch-Ups

There are many small-molecule drugs and supplements that have been promoted by suppliers and adopted by health-conscious individuals and longevity self-experimenters. Relatively simple longevity interventions, such as small-molecule drugs taken orally, do have definite value if they can slow the progress of aging and prolong health into old age. However, none of them can be expected to produce beneficial effects that persist for more than a few months or to have effects large enough to significantly extend a human lifespan.

6.1 Small-Molecule Supplements

6.1.1 Rapamycin

First among the small-molecule longevity interventions is *rapamycin*, an immunosuppressant drug often prescribed to prevent organ transplant rejection. Rapamycin (trade names *Sirolimus, Rapamune,* and *Hyftor*) is a protein product of the bacterium *Streptomyces hygroscopicus* and was isolated for the first time in 1972 from samples found on Easter Island. It was originally named "rapamycin" for the native name of the island, Rapa Nui. It was initially developed as an anti-fungal agent, but that use was abandoned when it was found to have strong suppression properties on the human immune system.

Suppression of mTOR (the mammalian target of rapamycin), specifically mTORC1, was first shown to be important as a longevity intervention in

J. G. Cramer, *How to Live Much Longer*, Copernicus Books,
https://doi.org/10.1007/978-3-032-17741-4_6

2003. Since then, it has been found to inhibit and slow aging in worms, yeast, and flies, and to improve the condition of various diseases of aging in mouse models. Rapamycin was first shown to extend lifespan in wild-type mice in a study published by NIH investigators in 2009; the studies have since been replicated in mice of many different genetic backgrounds. The results are further supported by the finding that genetically modified mice with impaired mTORC1 signaling actually live longer. There are also ongoing tests of the longevity effects of rapamycin on dogs that show preliminary positive results.

Rapamycin has potential for widespread use as a longevity-promoting drug [Bi16, Ha09, Ha24, Ka23, Le24, Ma14, Ma23], with evidence pointing to its ability to prevent age-associated decline of cognitive and physical health. This has led to many in the longevity community to undertake self-experiments with rapamycin, taking oral 5–7 mg doses weekly or biweekly. However, there are side effects to be considered. The reason for taking it in only 1–2 week intervals is because one wants to strike a delicate balance between maintaining a beneficial level of TORC1 suppression without allowing an undesirable rapamycin suppression of MTORC2. It is found that weekly or biweekly dosing of 5-10 mg of rapamycin primarily hits mTORC1, leaving mTORC2 largely intact.

Ultimately, because of the immunosuppressive side effects of rapamycin, more research is needed before it can be widely prescribed for longevity by the medical profession. For example, there are anecdotal reports of the incidence of the viral disease shingles afflicting individuals who had been taking weekly doses of rapamycin as a longevity intervention.

6.1.2 Metformin

Another small-molecule longevity intervention is metformin, a low-cost prescription drug commonly used to treat type 2 diabetes. Metformin has anti-inflammatory effects that should help protect against age-related diseases. It may also stimulate bone formation and reduce resorption, which could help maintain bone density and strength. Metformin is also thought to stimulate autophagy, which is the cellular cleanup process, providing removal of damaged proteins, mitochondria, and cells, potentially slowing aging and improving mitochondrial health.

Human subjects taking metformin have already statistically demonstrated its longevity extending potential. However, most of this research on metformin in humans has only included subjects with type-2 diabetes or pre-diabetes, so is not well established that it can also benefit non-diabetics.

Metformin can have mild side effects, including nausea, stomach upset, or diarrhea. More serious side effects, such as kidney damage, are also possible but rare. Its potential as a longevity intervention has led to many in the longevity community to self-experiment with metformin, taking 500 to 2,000 mg per day orally.

6.1.3 NAD+ Boosting: NMN and NR, Plus Apigenin

Sirtuins are a very important family of seven proteins that regulate many cellular processes, including DNA expression, repair, and control. However, they function only when well supplied with NAD+ (nicotinamide adenine dinucleotide), a coenzyme that is present in all living cells. By middle age, our NAD+ levels have dropped to half that of our youth. Studies have shown that boosting NAD+ levels increases insulin sensitivity, reverses mitochondrial dysfunction, and extends lifespan. NAD+ levels can be increased, at least in principle, by activating enzymes that stimulate the synthesis of NAD+, by inhibiting an enzyme (CD38) that degrades NAD+, and by dietary supplementation with NAD precursors, including *nicotinamide riboside* (NR) and *nicotinamide mononucleotide* (NMN).

Interestingly, a group recently found [Ho24] that subcellular NAD+ pools are interconnected, with mitochondria acting as a regulator to maintain NAD+ levels in the case of excessive consumption. They found that the age-related NAD+ deficiency was well tolerated unless mitochondria were damaged so badly that they could not perform this regulatory function.

So, does taking supplements with NAD+ precursors actually restore youthful NAD+ levels? Work by the Chini Group at the Mayo Clinic [Ca16] has shown that the action of the enzyme CD38, which increases with age, blocks the action of oral supplements like NR and NMN, intended to provide precursors of NAD+. Therefore, the best strategy for boosting NAD+ should include active suppression of the CD38 enzyme itself as well as such supplementation. Fortunately, the flavonoid *apigenin,* which is known to suppress CD38, is readily available as an inexpensive OTC supplement.

Note that because apigenin is a flavonoid, it has a very low solubility in water, leading to a low bioavailability. This problem can be addressed either by taking a large dose (~200 mg) or by taking liposomal apigenin at a smaller dose level. So, should one combine apigenin with NMN or with NR? NMN is only one reaction step away from NAD+ and perhaps is more efficient at producing NAD+, but it is relatively unstable. While NR is two reaction steps away from NAD+, it has been more widely tested, and it is more stable. Further, there are indications that it is synergistic with apigenin. An apigenin/

NMN combination might be better for muscular, metabolic, and mitochondrial support. On the other hand, the apigenin/NR combination, perhaps less efficient, might be better for boosting brain and liver NAD+ levels. We note also that over-the-counter NMN is rather expensive, about twice the cost of NR. I should add that there is some indication that even with liposomal apigenin, the blood concentration may not reach the level needed for significant CD38 suppression.

6.1.4 Klotho and GABA

Klotho, named after the Greek goddess who spins the thread of human life, is an enzyme that plays an important role in health. In humans, klotho is encoded by the KL gene. Subfamilies of klotho are highly expressed in the brain, liver, kidney, and skin. Although the majority of research has explored the effects of klotho's absence, it has been demonstrated by genetic modification that klotho overexpression in mice extended their average life span between 19 and 31% compared to unmodified normal mice. A beneficial mutation in the klotho gene (SNP Rs9536314) is associated with both life extension and increased cognition in human populations and in mice, but, curiously, is effective only if the gene expression is *not* present in every cell (heterozygosity). The cognitive benefits of klotho are primarily seen in later life.

Klotho is an antagonist of the Wnt signaling pathway (a crucial molecular pathway involved in various cellular processes, including cell fate determination, migration, and apoptosis). Chronic Wnt stimulation can lead to stem cell depletion and aging. Klotho inhibition of Wnt signaling can also inhibit cancer. The longevity effects of klotho are also a consequence of increased resistance to inflammation and oxidative stress.

Research has shown that extracellular vesicles (EV) of young mice carried more copies of klotho-producing mRNA than those of old mice. Transfusing such young EVs into older mice helped rebuild their muscles. The presence of senescent cells decreases α-klotho levels. Senolytics removes such cells and allows α-klotho levels to increase.

A research group in Toronto has found [Pr17] that the production of klotho is stimulated by the supplement GABA (*gamma-aminobutyric acid)*, which plays a role as a brain neurotransmitter and helps to regulate nerve activity and to maintain a balance between excitement and relaxation. GABA is a readily available over-the-counter supplement. The life-extension potential of GABA has led to many in the longevity community to take around 100–200 mg per day orally.

6.1.5 Resveratrol

"The French Paradox" is the 1980s conundrum that, despite a typical diet high in saturated fat and other rich foods, the French people appear to have lower rates of coronary heart disease. One 1980s theory asserted that the paradox could be explained by the high French consumption of red wine, which contains *resveratrol*, a compound that is reputed to have anti-hypertensive effects and to help relax blood vessels. Studies indeed show that resveratrol has antioxidant and anti-inflammatory properties, suggesting that it may be an effective longevity supplement, relieving oxidative stress and inflammation, improving cell function, and regulating cell aging and death.

However, much of the initial enthusiasm about resveratrol as an anti-aging agent dissipated somewhat in the 2000s, following test results indicating that health improvements in humans from regular resveratrol dosage are moderate at best. Nevertheless, research suggests that resveratrol's antioxidant properties do help strengthen the gut microbiome, decrease the risk of tissue damage, improve mood, and increase heart strength. Resveratrol is a polyphenol with low water solubility and bioavailability, which can be corrected by liposomal encapsulation or other measures. The life-extension potential of resveratrol has led many in the longevity community to take around 250–500 mg per day orally.

6.2 Mitophagy Boosters

Mitophagy is a specialized form of autophagy (cell cleanup self-removal) that performs the selective disassembly and disposal of damaged or dysfunctional mitochondria [Li23]. It is a natural cellular quality control mechanism to maintain cellular health and to ensure well-functioning mitochondria for robust ATP energy production. When a damaged mitochondrion shows a low intermembrane proton potential (becomes "low-voltage" or depolarized), the protein PINK1 accumulates on its outer surface. Parkin, attracted by PINK1, then moves to the damaged mitochondrion, signals that it is damaged, and summons the mitophagy bio-machinery to remove it. By this mechanism, illustrated in Fig. 6.1, damaged mitochondria with low inter-membrane potentials that are incapable of efficient ATP production are removed, to be replaced by the mitochondrial fission doubling to produce more healthy mitochondria.

Several popular small-molecule supplements are known to be mitophagy boosters: Urolithin A, Spermidine, Curcumin, Quercetin, and Alpha-Lipoic

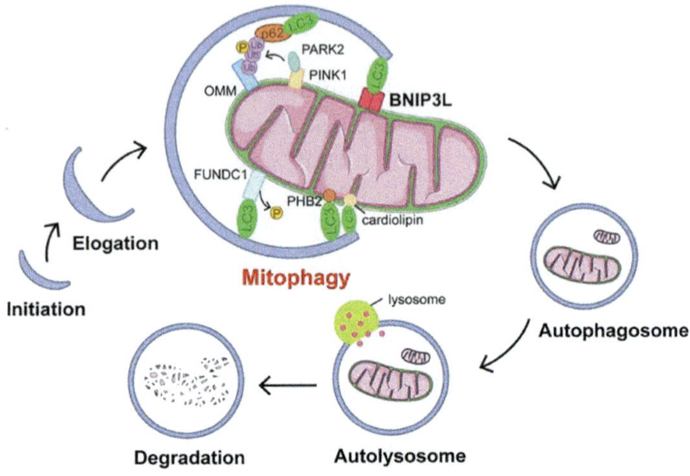

Fig. 6.1 The process of mitophagy. *Credit* Reference [Li23], used with permission

Acid. *Urolithin A* is a postbiotic metabolite produced by gut bacteria from chemical *ellagitannins*, which are found in pomegranates, walnuts, berries, and some other fruits. It is a potent mitophagy inducer, believed to act by activating the PINK1/Parkin mitophagy process described above.

Spermidine is a naturally occurring polyamine that is present in aged cheese, mushrooms, legumes, whole grains, and fermented products. It's also internally produced in humans. It has been shown to induce autophagy and mitophagy.

Curcumin is the active flavonoid compound in turmeric spice. It possesses potent antioxidant and anti-inflammatory properties and has been shown to induce mitophagy through various mechanisms, including modulating the Akt/mTOR pathway (a key intracellular signaling pathway that plays a crucial role in cell growth, proliferation, survival, and metabolism) and enhancing the PINK1-Parkin pathway.

Quercetin is a flavonoid found in many fruits, vegetables, and grains. It is a senolytic compound, and it has also been shown to induce mitophagy and to improve mitochondrial function.

Alpha-Lipoic Acid is a powerful antioxidant naturally produced in the body and found in foods like spinach, broccoli, and yeast. While primarily known for its antioxidant properties and role in mitochondrial metabolism, it has also been implicated in supporting mitochondrial health and potentially influencing mitophagy, often by improving mitochondrial integrity.

6.3 Heat and Cold Shock Proteins

Historically, human cultures around the world have developed traditions of immersing their bodies in very hot or very cold water. The affinity for hot baths goes back to the Roman Empire and beyond. Today, it includes Japanese hot tubs, Scandinavian saunas, Turkish baths, and European hot springs. At the other end of the temperature scale, there are many traditions of "polar bear clubs" with participants who plunge into icy waters. The traditional Scandinavian follow-up to a hot sauna is often a cold plunge. However, this practice is now somewhat discouraged because it is likely to over-stress the circulatory system.

As it turns out, there are good biological reasons behind both of these traditional practices: stimulating the production in the body of *heat shock* and *cold shock proteins*, some of which are associated with health and longer lifespan. Here are some of the most beneficial ones:

Heat Shock Proteins: *HSP70* prevents buildup of misfolded proteins and assists in protein refolding. *HSP90* stabilizes steroid hormone receptors, kinases that transfer phosphates from ATP, and transcription factors that copy gene coding to messenger RNA. *HSP27* reduces oxidative stress, improves muscle recovery, and slows neural degeneration. *HSP60* is a mitochondrial chaperone that is essential for mitochondrial protein folding and transport. *HSP32* is an anti-inflammatory that protects cell cytoplasm, reduces oxidative stress, and protects against cardiovascular and metabolic diseases.

Cold Shock Proteins: *RBM3* enhances nerve synapse plasticity and is strongly neuroprotective, supports recovery from hypothermia. *CIRBP* is an RNA chaperone that binds and stabilizes specific messenger RNAs during cold and other forms of cellular stress. (The long-lived bowhead whale's genome has been found to massively over-express CIRBP.) *YBX1* is involved in DNA/RNA regulation during stress, supports DNA repair, and also supports cell replication.

Thus, there may be significant longevity benefits in taking a bath while submerging in water as hot as is tolerable, staying in the hot water for 10 min or more, and possibly following that with a cold plunge or shower.

6.4 Vitamins, Minerals, and Supplements

The role of vitamins, minerals, and supplements as agents of longevity has long been over-hyped, in part because there has been a large and well-funded industry that, since the 1950s, has been aggressively promoting the need for them. The problem is that our normal foods already contain most of the vitamins and nutrients needed for good health, so accurately identifying areas that are in need of supplementation is difficult and controversial. However, there are a few of these items that, if taken regularly, are likely to make a positive contribution to these goals. The list includes vitamins B12, C, D3, and E, and coenzyme Q10, preferably in a liposomal form.

6.5 GDF11 Supplementation

GDF11 (Growth Differentiation Factor 11), is a rather large molecule with a molecular weight of 45,091 Da that acts as a growth hormone. Its supplementation has been tested on mice and shown to have some capacity to restore aging muscles, hearts, and brains. Recently, human experimenters have been self-testing by administering very small ($\sim 10^{-6}$ mg/kg) periodic injections of GDF11. The reported results have been impressive, with anecdotal reports of large improvements in many key biomarkers. This work implies is that weekly very small subcutaneous injections of GDF11 are an intervention that can reverse some of the effects of human aging, at least for a period of a few months.

6.6 HBOT Sessions for ATP Boost and Repair Stimulation

As we have been emphasizing in this book, the body's mitochondria use essentially all of our intake of oxygen in the metabolic chain, consuming it to produce ATP for cell energy. Therefore, it might be reasonable to suppose that if we simply breathed 100% pure oxygen (instead of normal air containing 21% oxygen) we might boost our ATP production, with many beneficial results. As it turns out, that seemingly plausible scheme is an oversimplification that has its own problems and side effects.

There are two channels for transporting oxygen from the lungs, where it is collected, to the mitochondria, where it is consumed: (1) oxygen carried by hemoglobin in red blood cells and (2) oxygen dissolved in the liquid blood plasma itself. In normal breathing of air at standard atmospheric pressure (1 ATM), the hemoglobin is already 98% saturated with oxygen, so its oxygen content cannot go much higher. Therefore, the only effective way of transporting significantly more oxygen to the mitochondria is to boost the amount of oxygen dissolved in the blood plasma. For breathing air at 1 ATM, the plasma transports 0.3 ml of oxygen per 100 ml of plasma, which represents only about 2% of the total delivered oxygen. Breathing pure oxygen does indeed raise the oxygen dissolved in plasma to about 2 ml per 100 ml, but it also increases the production of ROS and has other unwanted effects, leading to a set of unwanted symptoms called *oxygen toxicity*.

However, we can alternatively increase the amount of dissolved oxygen by increasing the ambient atmospheric pressure. It turns out that this also avoids increasing ROS or triggering oxygen toxicity. Breathing pure oxygen at 1.5 ATM raises the plasma-dissolved oxygen to about 3.2 ml per 100 ml, while increasing the pressure to 2 ATM raises it to about 4.4 ml per 100 ml. In other words, by breathing pressurized pure oxygen, we have a way of transporting more dissolved oxygen to our mitochondria by a factor of 11 (at 1.5 ATM) or 15 (at 2 ATM), with no accompanying ROS boost or oxygen toxicity.

Commercial HBOT (hyperbaric oxygen therapy) systems are designed to do just this. Essentially, an HBOT chamber is a metal tank or a heavy flexible plastic tube containing a pillow and mattress, which is closed with pressure seals and is equipped with a mechanical pump that can raise the internal pressure to 1.5 or 2.0 ATM. A subject is given an oxygen mask from which pure oxygen flows and is sealed into the chamber, which is then pressurized with ambient air. The subject rests inside for an hour or so, breathing pure oxygen at the elevated pressure while reading, playing digital games, scanning the Internet, working by phone or laptop, or napping. Commercial HBOT units that go to 2 ATM require a metal pressure tank and are fairly expensive (~ $30 k), but at this writing, there are plastic "home HBOT" units that operate at up to 1.5 ATM and are advertised on the Internet to cost around $2,800. We note that HBOT pressurization with pure oxygen would be expensive, would pose a severe fire danger, and is to be avoided.

What are the measured benefits of such an intervention? One HBOT study involving 35 healthy adults of ages 64+, each undergoing daily 60-min HBOT sessions at 2 ATM over 90 days, reported that analysis showed that blood immune-cell telomere length had increased by 20–38% and senescent immune cells had decreased by 11–37%, depending on cell type and individual. I have recently done 13 hours at 1.5 ATM of HBOT therapy, and I can personally report that I felt very boosted in energy for the next month or so.

6.7 Plasma Dilution and Young Plasma Replacement

There are two interventions involving blood plasma (the yellow liquid that remains after all the blood cells are removed from human blood) that are gaining popularity in the anti-aging community: plasma dilution and young plasma replacement. The Conboy group at U. C. Berkeley originated the plasma dilution concept (although it bears a striking resemblance to the standard procedures of plasma donation used by many blood banks). The Conboy's basic assumption is that the blood of the elderly contains unwanted agents that promote the symptoms of aging, and that by replacing old blood plasma with normal saline solution containing albumin, one can eliminate these undesirable components, leading to beneficial effects [Me20]. This intervention does indeed appear to provide significant temporary benefits, supporting their "unwanted agents" hypothesis. The Conboy group's test method of estimating biological age from the variance of protein expression, described above, indicates a biological age reduction by several years following completion of sessions of plasma dilution therapy.

The other intervention of note involves soliciting plasma donations [Al25] from young paid volunteers (usually college students) under the age of 22. Typically, no donor-to-recipient matching is done except male-to-male or female-to-female. The extracted "young plasma" is then infused into the bloodstreams of elderly recipients. Reportedly, this procedure costs the recipients about $10 k per session. Tests of the results of such interventions indicate significant temporary benefits that seem to be somewhat better than those observed for plasma dilution therapy.

The question then is, *what is contained in the young plasma that produces these added benefits to elderly recipients?* One likely possibility is that the plasma contains many exosomes that bring with them many beneficial agents. Exosomes are membrane-walled vesicles about 40–160 nm in diameter that are found in blood and that transfer their contents to the cells they encounter by wall absorption. They are known to contain a variety of components, including mitochondria, proteins, lipids, nucleic acids, DNA fragments, and metabolites. There are also mitochondria in the plasma itself. There is a strong possibility that the transferred mitochondria from these exosomes and plasma are among the agents that produce the observed beneficial effects. Essentially, young plasma replacement may be a roundabout way of doing a small-volume mitochondrial transplantation, but with no haplogroup matching (see Secs. 6.10 and 10.4 below).

6.8 Platelet and PRP Donations to Boost Mitochondria

The platelets are the blood's injury first-responders that induce blood coagulation at wound sites. Each platelet contains 5–9 mitochondria. The platelets in the blood are continually produced by the bone marrow. Platelets have a lifespan of only 8–12 days, after which they expel their mitochondria in extracellular vesicles and are removed by the immune system. The mitochondrial-containing vesicles are then absorbed by nearby cells, particularly the immune cells. Therefore, one possible way to do small-volume mitochondrial transplantation is to infuse the recipient with donated platelets containing healthy mitochondria, preferably haplogroup-matched. Within a week or so after their infusion, the donated platelets will provide the immune and other cells with their mitochondrial contents. There is also a proprietary method of extract mitochondria-containing "mitlet" vesicles from the platelets and injecting these, delivering the mitochondria immediately.

There is some evidence, from animal studies involving fluorescent mitochondria, that the body has a system of mitochondrial distribution that moves mitochondria from sites where there is a surplus to sites where they are needed. Therefore, this platelet form of mitochondrial transplantation should provide general ATP boost benefits [Gr23]. In particular, platelet donation from a maternal-line relative to a child suffering from a mitochondrial genetic disease may provide an easy and effective method of treatment. The quantity of delivered mitochondria will necessarily be small, but, as suggested by the young plasma replacement results described above, they should have definite beneficial effects.

The established and widely available blood treatments with platelet-rich plasma (PRP) may be considered similarly useful [Be22]. Normally, the platelets for PRP come from the subject's own blood and are injected into localized areas in need of therapy. However, if longevity is the objective through mitochondrial transplantation, the platelets should come from a younger donor, preferably a young maternal-line relative with a matching mitochondrial haplogroup, and should be infused directly into the bloodstream for whole body distribution.

6.9 Stem Cell and Stem-Cell-Derived Exosome Therapies

Stem cells are the body's active mechanics, ready to repair or replace damaged cells of all types, while also manufacturing and delivering components as needed [Sz21]. If the body's supply of stem cells becomes damaged or depleted by age or disease, or if the available stem cells have significant DNA damage, the overall health of the body declines.

There are several existing and widely used stem cell transplant and enhancement procedures. The most common medical procedure in use is the bone-marrow transplant, in which stem cells removed from donor bone marrow are used to treat patients with blood cancers such as leukemia. This procedure is done by clearing harmful cancer cells in the patient's bone marrow using ionizing radiation and chemotherapy, and then transplanting bone marrow containing healthy stem cells obtained from a related or matching donor.

There are also a significant number of other stem cell transplant techniques now in testing or near FDA approval in the US, including prospective treatments for Alzheimer's, retinal or corneal regeneration, joint restoration, wound healing, etc. Outside the US, some nations with more permissive and flexible health treatment regulations (e.g., Panama, South Korea, Japan, Mexico, India, …) have clinics offering quite extensive stem cell procedures and therapies.

Where do the stem cells used for transplants come from? If patients happen to have their own original frozen placenta or cord blood (unlikely), the perinatal stem cells from their own body can be used. An alternative stem cell source is to use the patients' current body tissue (abdominal fat, muscle tissue, blood stem cells, extracted bone marrow, …). The extracted tissue sample is usually epigenetically deprogrammed externally to convert the specialized cells to stem cells, which may then be used directly or may be multiplied in an external bioreactor. The resulting pluripotent stem cells are infused back into the patient's body by local injection to damaged areas or by slow infusion into the bloodstream to distribute them more generally.

Another alternative source is to obtain stem cells from an unmatched donor. Offshore stem cell clinics are known to purchase birth-derived placentas from local hospitals and from female donors who recently gave birth, so that they can extract the perinatal stem cells from them. These are reputed to have greater regenerative capacity than stem cells from other sources. The perinatal stem cells are then processed as described above and administered to the patient. In this case, the nuclear DNA and mtDNA of the cells and mitochondria that the patient receives will likely differ from those of existing cells, creating a condition called *heteroplasmy*. It is not known if this condition produces any serious downstream problems for the recipient, but there are hints from mouse and human studies that it might.

There is another potential negative side effect from stem cell enhancement procedures. Particularly in pathological states like inflammation, stem cells may use various mechanisms, such as active transport or changes in the permeability of the blood-brain barrier, to reach areas that had not been targeted for treatment. Also, in some cases, damage can arise in stem cell transplantation if there is any contamination, if the dose contains old or genetically damaged cells, if the stem cells are not well separated from other tissue material, or if some fraction is of an unwanted cell type. The immune system also may react to injected stem cells in an uncontrolled way. Further, most stem cell therapies carry some risk of inducing tumor formation from random malignant stem cells.

A new alternative, *stem-cell-derived exosomes,* has recently been introduced. Exosomes, as described above, are membrane-walled vesicles 40–160 nm in diameter that can be released from stem cells. They can then deliver their contents to other cells. Exosomes contain a variety of components, including mitochondria, proteins, lipids, nucleic acids, and metabolites, giving them therapeutic value. They are normally non-toxic, do not trigger the immune system, and cannot form tumors. They probably play a role in wound healing and tissue repair. Animal studies with rats [Sh23] have shown that significant reductions in epigenetic age can be achieved by administering young-blood-derived extracellular vesicles, a form of exosome.

There have also been human clinical trials investigating exosome-based therapies for a few diseases. One of these has used stem-cell-derived exosomes for respiratory therapy. The results suggest that stem-cell-derived exosomes provide much the same effects as their donor stem cells, while avoiding several drawbacks of the latter. For example, exosomes cannot self-replicate, eliminating the possibility of tumor formation. They are stable enough for long-term frozen storage or, after freeze-drying, room-temperature storage. Their small size facilitates sterilization by filtration. They can be administered by several routes, including injection and nebulization (for inhalation to treat lung disease).

There remains an unresolved question: *what does stem cell therapy actually do, and how does it work?* The simplistic answer is that the stem cells are new replacements for old and damaged cells. However, this view is currently placed in doubt by the observation that exosome-based therapies confer much the same benefits [Ta24], and yet exosomes *cannot* transform into new cells for replacement.

One possible alternative explanation is that stem-cell/exosome therapy is a way of supplying damaged cells with new and healthy mitochondria. The average stem cell contains between 50 and 500 mitochondria, and when they are processed to produce exosomes, the mitochondria go with them. Injected stem cells or exosomes then release their mitochondrial load, which the body distributes to cells and organs in need. Thus, at least hypothetically, stem-cell/exosome therapy can be viewed as a form of small-volume mitochondrial transplantation.

6.10 The Mitochondrial Haplogroup Problem

As we have seen, many of the interventions discussed in this chapter may be in effect the delivery of new mitochondria to cells in need. In many cases, the mitochondria come from an unrelated donor whose mtDNA is very likely to have a different inherited single-point mutation pattern or *mitochondrial haplogroup*. The potential problems from such mismatching will be discussed in Chap. 10, Section 10.4.

In present practice for interventions like young-plasma replacement and stem cell therapy, usually no effort at all is made to measure or match donor-recipient mitochondrial haplogroups, so mtDNA mismatches and limited heteroplasmy are the usual situation. What quantity of haplogroup-mismatched mitochondria can be transplanted before downstream problems are likely to arise? The problem is that in a cell the mother's mitochondria-related nuclear DNA and mtDNA of the recipient have co-evolved over many millennia and "fit together", while the transplanted mtDNA and nuclear DNA may not co-exist without incompatibility problems. What and how serious are those problems? We simply do not know. At a guess, such problems might appear when the mismatched mtDNA fraction reaches around 30%.

6.11 The Bottom Line: Is a 5% Solution Sufficient?

In using longevity interventions, one is tempted to think that if one effective intervention reduces your biological age by two years, then using 10 such interventions in succession should reduce your biological age by 20 years. It doesn't work that way.

The basic problem with all of the longevity interventions discussed above is that their net effects are relatively small (~ 5%) and the error bars are large, so any measured improvements have large uncertainties (20–50%), and their duration of effectiveness is relatively short (weeks or months). They may improve some biomarkers, and they may reduce some measure of biological age by a few years, but they do not fix the basic problem, which is that our mtDNA is suffering increasing damage and as we age, our supply of ATP energy is dropping progressively and exponentially from a healthy level to near zero.

To me, at least, it is clear that to deal with the fundamental problem of human aging, much stronger measures are required. We need to be able to extend human life by more than just a few years. The next two chapters will consider some of these measures. Table 6.1summarizes what we have discussed in this chapter.

Table 6.1 Short-term longevity interventions

Intervention category	Specific method	Description and procedure	Proposed benefits and reported effects	Key considerations and limitations
Thermal stress	Heat therapy (e.g., hot baths, saunas)	Submerging in hot water (as hot as tolerable) for 10+ min	Stimulates production of heat shock proteins (HSPs: 70, 90, 27, 60, 32). Benefits include: preventing protein misfolding, reducing oxidative stress, anti-inflammatory effects, and supporting mitochondrial function	Traditional practice with a biological basis. Following with a cold plunge is now discouraged due to potential circulatory stress
	Cold therapy (e.g., cold showers)	Exposure to cold water for short periods	Stimulates production of cold shock Proteins (RBM3, CIRBP, YBX1). Benefits include: neuroprotection, enhanced nerve synapse plasticity, RNA stabilization, and support for DNA repair and cell replication	Often used in conjunction with, but not immediately following, heat therapy
Supplements	Vitamins and CoQ10	Regular supplementation with vitamins B12, C, D3, E, and coenzyme Q10 (preferably in liposomal form)	Aims to address potential deficiencies and support general cellular health and energy production	Role is often over-promoted; normal diet provides most needs. Identifying true deficiencies is difficult
	GDF11 injections	Weekly, very small subcutaneous injections (~ 10^{-6} mg/kg)	Anecdotal reports of large improvements in key biomarkers. Shown in mice to restore aging muscles, hearts, and brains	Human evidence is largely anecdotal. Involves injections of a large molecule (45 kDa). Effects may last only a few months

(continued)

Table 6.1 (continued)

Intervention category	Specific method	Description and procedure	Proposed benefits and reported effects	Key considerations and limitations
Oxygen therapy	Hyperbaric oxygen therapy (HBOT)	Breathing 100% pure oxygen in a pressurized chamber (1.5–2.0 ATM) for ~ 60 min per session.	Study reported: Telomere lengthening (20–38%), reduction in senescent cells (11–37%). Anecdotal reports of significant energy boost	High-pressure (2 ATM) units are expensive (~ $30 k). Lower-pressure (1.5 ATM) "home" units are cheaper (~ $2.8 k). Avoids ROS/ toxicity of normobaric pure oxygen
Blood-based therapies	Plasma dilution	Removing a portion of the patient's blood plasma and replacing it with a saline-albumin solution	Based on the hypothesis that old plasma contains pro-aging factors. Conboy group data suggest a reduction in biological age (measured by protein expression variance)	Similar to standard plasma donation. Benefits are reported to be significant but temporary
	Young plasma replacement	Infusing plasma from young donors (under 22) into an elderly recipient	Reported to produce significant temporary benefits, seemingly better than plasma dilution. Cost is ~$20 k per session	Proposed mechanism: Beneficial contents of young plasma, such as exosomes containing healthy mitochondria, proteins, etc. no donor matching is typically done
	Platelet/PRP infusion	Infusing platelets from a young donor into the recipient's bloodstream. Platelets contain 5–9 mitochondria each	Proposed to be a form of small-volume mitochondrial transplantation. Platelets release their mitochondria in vesicles absorbed by other cells, providing a general ATP boost	For longevity (vs. localized repair), platelets should come from a young donor, ideally a maternal-line relative for haplogroup matching

Intervention category	Specific method	Description and procedure	Proposed benefits and reported effects	Key considerations and limitations
Cell and vesicle therapy	Stem cell therapy	Harvesting stem cells (from patient or donor), reprogramming/multiplying them, and infusing them back (locally or systemically)	Aims to repair or replace damaged cells. Used for conditions like leukemia, with other applications (Alzheimer's, joint repair) in testing	Risks include: Immune reaction, tumor formation, contamination, and heteroplasmy (if from a donor). Invasive and expensive
	Stem-cell-derived exosomes	Administering exosomes (40–160 nm vesicles) derived from stem cells, containing mitochondria, proteins, lipids, nucleic acids	Reported to provide benefits similar to stem cell therapy (e.g. reduced epigenetic age in rat studies) but without cells. Advantages: non-immunogenic, cannot form tumors, stable for storage, can be sterilized	Proposed mechanism may be the delivery of healthy mitochondria and other beneficial cargo. Seen as a potentially safer alternative to whole stem cell therapy

References

[Al25] Alkahest Inc., 125 Shoreway Road, Suite D San Carlos, CA 94070; https://alkahest.com.

[Be22] T. Benson, S. Patel, B. C. Albensi, V. B. Mahajan, A. Adlimoghaddam, and H. Saito, "Platelet extracellular vesicles and their mitochondria content improve the mitochondrial bioenergetics of cellular immune recipients," ***Transfusion 2023***, 1–14 (2023); https://www.biorxiv.org/content/10.1101/2023.09.08.556161v1.

[Bi16] A. Bitto, et al, "Transient rapamycin treatment can increase lifespan and healthspan in middle-aged mice," ***eLife*** (2016); 10.7554/eLife.16351.

[Ca16] J. Camacho-Pereira, M. G. Tarragó, C. C. S. Chini, V. Nin, C. Escande, G. M Warner, A. S Puranik, R. A. Schoon, J. M. Reid, A. Galina, E. N. Chini, "CD38 Dictates Age-Related NAD Decline and Mitochondrial Dysfunction through an SIRT3-Dependent Mechanism," ***Cell Metabolism 23*** (6), 1127–1139 (2016); https://doi.org/10.1016/j.cmet.2016.05.006.

[Gr23] A. Grichine, S. Jacob, A. Eckly, J. Villaret, C. Joubert, Appaix, F., Pezet, M., Ribba, A.-S., Denarier, E., Mazzega, J.-Y., Rinckel, J., Lafanechère, L., & Elena-Herrmann, B. "The fate of mitochondria during platelet activation," ***Blood Advances 7***(20), 6290–6302 (2023).

[Ha09] D. E. Harrison, *et al,* "Rapamycin fed late in life extends lifespan in genetically heterogeneous mice," ***Nature 460*** (7253), 392–395 (2009).

[Ha24] G. Harinath, et al, "Safety and efficacy of rapamycin on healthspan metrics after one year: PEARL Trial Results," medRxiv (2024); https://doi.org/10.1101/2024.08.21.24312372.

[Ho24] L. E. Høyland, M. R. VanLinden, M. Niere, Ø. Strømland, S. Sharma, J. Dietze, I. Tolås, E. Lucena, E. Bifulco, L. J. Sverkeli, C. Cimadamore-Werthein, H. Ashrafi, K. F. Haukanes, B. van der Hoeven, C. Dölle, C. Davidsen, I. K. N. Pettersen, K. J. Tronstad, S. A. Mjøs, F. Hayat, M. V. Makarov, M. E. Migaud, I. Heiland, and M. Ziegler, "Subcellular NAD+ pools are interconnected and buffered by mitochondrial NAD+," ***Nature Metabolism 6***, 2337 (2024).

[Ka23] T. L. Kaeberlein, et al, "Evaluation of off-label rapamycin use to promote healthspan in 333 adults," ***Geroscience*** (2023).

[Le24] D. J. W. Jun Lee, et al, "Targeting aging with rapamycin and its derivatives in humans: a systematic review," ***Lancet Healthy Longev.*** (2024) 18.

[Li23] Y. Li, W. Zheng, Y. Lu, Y. Zheng, L. Pan, X. Wu, Y. Yuan, Z. Shen, S. Ma, X. Zhang, J. Wu, Z. Chen, and X. Zhang, "BNIP3L/NIX-mediated mitophagy: molecular mechanisms and implications for human disease," ***Cell Death and Disease 13*** (14), (2022); 10.1038/s41419-021-04469-y.

[Ma14] J. B. Mannick, et al. "mTOR inhibition improves immune function in the elderly," ***Sci Transl Med.*** (2014).

[Ma23] J. B. Mannick, et al, "Targeting the biology of aging with mTOR inhibitors,". ***Nature Aging 3***, 642–660 (2023); https://doi.org/10.1038/s43587-023-00416-y.

[Me20] Melod Mehdipour, *et al*, "Rejuvenation of three germ layers tissues by exchanging old blood plasma with saline-albumin", ***Aging (Albany NY) 12***: 8790–8819 (2020); https://doi.org/10.18632/aging.103418.

[Pr17] G. J. Prudhomme, Y. Glinka, M. Kurt, W. Liu, and Q. Wang, "The anti-aging protein Klotho is induced by GABA therapy and exerts protective and stimulatory effects on pancreatic beta cells," ***Biochemical and Biophysical Research Communications 493*** (4), 1542–47 (2017); https://doi.org/10.1016/j.bbrc.2017.10.029.

[Sh23] L. G. Shamagian, R. G. Rogers, K. Luther, D. Angert, A. Echavez, W. Liu, R. Middleton, T. Antes, J. Valle, M. Fourier, L. Sanchez, E. Jaghatspanyan, J. Mariscal, R. Zhang, and E. Marbán, "Rejuvenating effects of young extra-cellular vesicles in aged rats and in cellular models of human senescence," *Nature: Scientific Reports 13*, 12240 (2023); https://doi.org/10.1038/s41598-023-39370-5.

[Sz21] A. Szubarga, M. Kamińska, W. Kotlarz, S. Malewski, W. Zawada, M. Kuczma, M. Jeseta, and Antosik, "Human stem cells—sources, sourcing and in vitro methods," Medical Journal of Cell Biology 9(2), 73–85. (2021); https://doi.org/10.2478/acb-2021-0011.

[Ta24] Fei Tan, et al, "Clinical applications of stem cell-derived exosomes," ***Signal Transduction and Targeted Therapy 9***:17 (2024).

7

Longer-Term Interventions: Senolytics, Epigenetics, Telomeres

In addition to the small-molecule and other longevity interventions discussed in the previous chapter, which provide benefits for perhaps a few months, there are a few in-development or currently available longevity interventions that promise more long-term benefits. We will discuss these here and in the following chapter.

7.1 Senolytics

One key element in the process of aging, as discussed in section 5.9 above, is the *senescent cell*. As the body ages, many of its cells become damaged and unable to fulfill their intended functions [Cr24]. This damage may be the result of mutated mitochondria, over-short telomeres, transcription errors during cell division, ionizing radiation, oxidation, toxins, diseases, and a long list of other causes. Such damaged cells are either removed by autophagy or become senescent. In the latter case, they are blocked from any further cell division. They often cease their normal functions and behave badly, secreting harmful chemical signals (SASP) that degrade tissue function, generate inflammation, interfere with the performance of nearby cells, and induce neighboring cells to become senescent too. The inflammation associated with aging is largely the result of senescent cell accumulation. Organs that accumulate the most senescent cells tend to be those in damaging environments that replicate with the highest frequency, i.e., skin cells and stomach and intestinal linings.

J. G. Cramer, *How to Live Much Longer*, Copernicus Books, https://doi.org/10.1007/978-3-032-17741-4_7

In principle, the immune system should automatically detect and remove these senescent cells, but for unknown reasons it does not. Instead, senescent cells accumulate with age, and their effects show up in many ways: "papery" skin, eye lens cataracts, sarcopenia (muscle loss), poor recovery from exercise, arthritis, organ degradation, a general tendency to form cancers, amyloid buildup in tissues and organs, and so on.

One technique for identifying senescent cells exploits a natural mechanism that senses their damaged condition and causes them to express a protein called "p16", which has the function of turning off the cell division machinery. (The p16 protein is also known to suppress tumor growth and is sometimes called a "tumor suppressor protein").

In a 2016 experiment at the Mayo Clinic that was performed on transgenetic mice [Ba11, Ba16], researchers rewrote a tiny portion of the mouse genetic code, providing the subject mice with cells that would produce a drug-triggered "suicide" protein and be cleared by apoptosis whenever they attempted to produce protein p16. This modification gave the researchers a drug-activated "kill-switch" that, when turned on, would effectively scrub the bodies of middle-aged mice clean of all senescent cells. The result of this trial was that the subject mice, once their senescent cells had been removed, were conspicuously larger, more active, and healthier than a control group. In particular, the subject mice after treatment had their average lifespans increased by about 25% and showed a reduction in the incidence of cancer of about 50%.

7.1.1 Plasmids for Senolytics

The Oisin Bio bio-startup has used standard commercial DNA sequencing technology to sequence a special "plasmid", a small ring of DNA containing the coding for a p16-triggered promoter that controls the expression of an inducible suicide gene called "iCasp9". The iCasp9 gene produces a protein that initiates cell death, but that protein does not activate unless a small-molecule dimerizer is injected. The dimerizer causes the two halves of the iCasp9 protein to bind together, immediately triggering cell "apoptosis", i.e., the immediate disintegration and local cleanup of the target cell. When this plasmid is present, if the cell attempts to express the p16 protein it is removed by apoptosis.

However, there is a delivery problem because the plasmid must be placed within the cell's surrounding outer membrane in order to operate. This problem is solved by placing the plasmid in the interior of a "fusogenic lipid nanoparticle" (LNP), a spherical "bubble" formed from a double layer of waxy lipid molecules identical to those that form cell membrane walls. When a fusogenic LNP contacts a cell membrane, it fuses with it and delivers the enclosed material into the cell's interior.

When the plasmid-loaded LNPs are injected into the subject's bloodstream, they are quite "democratic" and non-specific. They do not seek out cells with any specific functions or surface characteristics. Rather, they go everywhere the bloodstream goes and deposit their genetic loads in any of the cells they encounter. After this LNP dispersal of the plasmids, a carefully controlled amount of dimerizer is injected into the bloodstream. The size of the dimerizer dose controls the number of senescent cells that will be cleared through apoptosis in a particular session. Tests of this technique with mice show roughly the same results as the Mayo Clinic tests described above with genetically modified mice.

There has already been a significant spin-off from this longevity technique. They are exploring the use of the same technique to target tumors that are known to express the cancer-related protein p53. A p53-triggered plasmid was synthesized, and a trial was performed on a strain of immunodeficient mice, chosen because the mouse immune system wouldn't reject the human PC3 prostate cancer cells implanted in their flanks. Surprisingly, the test showed as much as a 90% reduction in tumor mass in 24–48 h of treatment. These results were astonishing and virtually unprecedented. Human trials are now being planned in which the same treatment will be applied to males with late-stage prostate cancer.

7.1.2 D + Q + F Senolytics

There is also a simpler senolytic technique that is currently in use by many longevity "self-experimenters". It is the "D + Q + F" senolytic technique. It involves taking large oral doses of the small-molecule senolytic agents Dasatinib, Quercetin, and Fisetin on three consecutive days, with these sessions repeated every 3 or 4 months. A preliminary version of the technique was invented for mice by the Kirkland group at the Mayo Clinic. They demonstrated that D + Q works well on mice, extending health-spans and lifespans, and that D + Q infused in human cell cultures produces apoptosis in senescent cells [Sa21].

One of the small molecules used is the expensive proprietary anti-cancer drug *dasatinib* (D). The other two are the flavonoid *quercetin* (Q) and, more recently, the flavonoid *fisetin* (F). Fisetin and quercetin, as flavonoids, both have the problem of very low solubility in water. Therefore, if taken orally in powder or capsule form they have low bioavailability, resulting in most of the flavonoid being excreted rather than reaching the bloodstream and the target senescent cells. (The flavonoids apigenin, mentioned below as a NAD+ enhancer, and curcumin, reputed to have longevity properties, both have similar solubility problems.)

Fortunately, there are new techniques for encapsulating these flavonoids in liposomes or other coatings, so that they are protected from digestive system breakdown and their availability greatly increased. The senolytic interventions D + Q and D + Q + F, with suitable bio-availability enhancements, are currently the most widely used senolytic techniques. However, they are not the only (or even the best) way of ultimately doing senolytics and senescent cell clearance.

We note that the well funded biotech startup Unity Bio [Un25] conducted human trials investigating the use of their proprietary small molecule senolytic agents in clearing senescent cells in athriitic joints and other localized areas. However, at this writing, after the widely publicized trial of this technique showed little improvement, the company ceased operation.

7.1.3 Senolytic Replacement, Energy, and Targeting, Problems

There are several problems with stand-alone senolytics: (1) not all of the organs of the human body have the same burden of senescent cells, (2) when senescent cells are forced into apoptosis, that may create a cell vacancy in the organ that is not immediately filled, and (3) it is energetically cheaper to operate a senescent cell for a day than to command it to go into apoptosis (by a factor of 150 or more). It also costs less energy in the short term to send the senescent cell into necrosis (which is immediately energy free but has large long-term cleanup energy costs). Therefore, if energy is in short supply, a senolytic intervention may send senescent cells into necrosis rather than apoptosis, putting necrotic toxins into the bloodstream. This may be the reason for reports from older D + Q + F self-experimenters that their senolytic sessions produce transient mild headaches, low fever, intestinal cramps, and flu symptoms for about a day.

There is also a problem with delivering the senolytic clearance-inducing molecules to their desired cell targets. Desatinib is reported to target mainly fat cells and those of the intestinal lining. The senolytic flavonoids quercetin and fisetin are reported to primarily target skin cells and cells of the blood vessel linings. It is not clear to what degree the senescent cells in other major organs (heart, kidney, liver, pancreas, ...) are targeted and affected by these small molecule interventions.

7.2 Epigenetic Reprogramming

An individual, when he is 20 years old and when he is 80 years old, has exactly the same nuclear DNA except for a few accumulated mutations, yet he has a very different body and very different performance levels. Somehow, the cells of the 80 year-old "know" that they are old cells, independent of their DNA-supplied instructions. The difference is that the *epigenetic programming*, the pattern of methyl radicals attached to the CpG sites of the DNA, has been changed with age, altering which of the DNA-encoded proteins are currently being produced and which are absent because their genes have been silenced.

This situation suggests that if somehow an aged individual's epigenetic programming could be reset to the former young methylation profile, the age-related changes would perhaps be erased, and the individual's youth and vigor might be restored. So, is there any way of "resetting" the epigenetic programming of aging cells to a younger state? The answer seems to be: *yes.*

We already know that such a reset is possible in principle. When a man and a woman, perhaps in their early 40 s, conceive a child, both parents contribute their DNA, and the mother contributes one of her reproductive cells (as an ovum) to the new infant. Somehow, this particular cell must have all of the mother's accumulated environmental damage erased and her 40 years of epigenetic programming reset to zero. The child starts life with new embryonic cells, not 40 year-old ones [Ke21]. How does this happen? There is now a body of research providing the apparent answer to this question: Nature has provided a set of special reset proteins that accomplish the epigenetic reprogramming task.

In Chap. 3, Sect. 3.1, we have already described the use of Yamanaka-tyoe reset proteins in erasing the epigenetic programming to become "brand new" pluripotent stem cells. As described there, Sarkar et al. [Sa20] used synthetic messenger RNA to stimulate the production of a six-protein cocktail of OCT4, SOX2, KLF4, c-MYC, LIN28, and NANOG (acronym "OSKMLN"). Using this technique, they discovered that cell cultures could be reprogrammed to be genetically younger, but to retain their original functional cell type, provided that the treatment process was halted after four days. They chose four days because that time was observed to be just before the reset "point of no return" was reached, after which the treated cells reverted to completely undifferentiated pluripotent cells. In other words, they found that *human cells can be rejuvenated*. The implication of this work is quantitative evidence that the epigenetic programming of human nuclear DNA can be reset, perhaps to reverse all the effects of aging.

So, how could this rejuvenation be made a real treatment for aged humans? How does one put selected proteins into cells for such treatments? To answer this question, let's discuss a bit of the mechanics of cell biology and look into how a cell actually produces a protein like those discussed above.

If a cell wants to produce a protein, it generates a transcription factor enzyme. The transcription factor is coded to find, within the library of the cell's DNA sequences, the particular "promoter region" (a unique identifying DNA base sequence) indicating that the instructions for producing the targeted protein follow in sequence. The transcription factor finds and attaches to the DNA's promoter region, then zips down the connected sequence of bases, producing a chain of messenger RNA (mRNA) containing the information. The mRNA that has been generated finds its way to a ribosome (a protein-assembly organelle located within the cell's cytoplasm) and threads through the ribosome like a punched paper tape, delivering the instructions and causing the ribosome to assemble the requested protein out of locally available amino acids. That protein is then released from the ribosome and does its job within or outside the cell.

What the Sarkar group did in demonstrating that human cells can be "reset" was to use the new laboratory bio-machinery of synthetic RNA sequencing and PCR to mass-produce mRNA sequences coded for each of the six reset proteins of the OSKMLN cocktail, using a process similar to that developed for the mRNA COVID vaccines. They then introduced this six-element cocktail of mRNA sequences into the cell interiors of their cell cultures, using several specialized commercial transfection compounds and procedures that had been optimized for the different cell types investigated. The ribosomes of the cell interiors then used the mRNA to generate the desired rejuvenation proteins, and the proteins then went to work reprogramming the cells to a more youthful state. The cell cultures "reprogrammed" by this procedure were then tested to determine their biological state. The researchers found that in the aging cells tested, even those obtained from 60–70 year-old humans, eight of the nine hallmarks of aging were reset to the state expected for young cells, with only telomere length (see below) remaining unchanged.

This test is a remarkable demonstration that it is possible in the laboratory Petri dish, and perhaps even in living organisms, to wipe away the damage of age and restore cells to a youthful state. The implication of this work is that DNA epigenetic reprogramming can be accomplished, which may reverse the effects of aging.

As mentioned in Chap. 3, Sect. 3.1 above, a promising alternative to the Yamanaka factors is a new reset protein called SB000, recently discovered by

Shift Biosciences in the UK [Sh25]. Tests show that SB000 resets the epigenetic programming of treated skin cells to a younger state without causing the treated cells to forget their function or become pluripotent [Ca25]. This development could make an epigenetic reset much simpler to implement.

It should be mentioned that there is also another method, virus delivery, for infusing aging cells with Yamanaka-type reset proteins. It has been widely used and reported in the literature, usually as applied to transgenetic mice. It involves synthetically constructing a DNA programming sequence that will produce a set of reset proteins on command when a particular exotic antibiotic, *doxycycline,* is present. A DNA "cassette" with the switchable reset genes is inserted into a nuclear DNA chromosome of the subject by an AAV (a "tamed" adenovirus used for DNA delivery). Then, during the tests, the doxycycline is administered with careful timing to provide bursts of the reset proteins, and the effects observed. However, there are ethical issues in applying this delivery technique to humans because it may permanently modify the subject's DNA. Such DNA modifications, at least if done to the genes of humans of breeding age, could, in principle, be passed on to offspring.

Are Yamanaka factors or SB000 the only way to achieve epigenetic reprogramming to a younger profile? Apparently not. In a parabiosis experiment [Co05, Co13, Zh23] in which a young mouse and an old mouse were strapped together and their bloodstreams connected for three months, the experimenters found that for three different cell age clocks, including Horvath CpG methylation, the blood and liver tissue of the older mouse showed that cells had been epigenetically reset to a younger profile. The blood cells decreased in age by 19–28% and liver cells by 5–26%. This rejuvenation was observed to persist for at least two months after the mice were separated.

What produced this reset? That is not clear, but other experiments have shown that exosomes derived from young blood produce similar age-reversal effects. This leads to the speculation that the epigenetic reprogramming observed in parabiosis is the result of exosome transfer of young mitochondria to the older mouse, boosting the ATP supply and signaling to the epigenetic-profile controls that genes silenced to save energy can be restarted.

However, for the oldest human subjects, there are also likely to be problems with all such epigenetic reprogramming efforts that do not change the cellular energy supply, independent of the delivery method. The Yamanaka-type epigenetic reprogramming process is not "free" in the energy sense. In fact, it is rather energy expensive [Ci24, Fo18], demanding an estimated 3–10 times as much ATP energy as is required for normal somatic cell operation. In embryos and young individuals, that energy demand is not a problem because healthy mitochondria are churning out ATP at top speed, and it is a time of energy abundance.

However, the body of a 70–90 year-old human is in the grip of a severe ATP energy shortage. There has already been much accumulated damage to the mitochondrial DNA, with ATP energy in short supply. Normal epigenetic programming, acting in "energy miser" mode, has already silenced many genes with large energy demands in favor of substitute genes with lower energy requirements. In this situation, using selected Yamanaka factors to *force* epigenetic reprogramming on a cell when there is not sufficient energy to complete the process could be a prescription for disaster. It is not clear what happens when such reprogramming is forced into action and then halted in midstream due to "starvation" when no energy is available. Bad outcomes are to be expected, and the goal of reprogramming aging cells may fail.

7.3 Restoring Telomere Length

In 1990, the *telomere* was discovered. A telomere in humans is a special DNA structure at the ends of chromosomes that has the repeating base pair sequence TTAGGG TTAGGG TTAGGG TTAGGG … As illustrated in Fig. 7.1 and indicated by the pink circles, the telomeres are the sequences found on both ends of all of the human chromosomes in the cell's nucleus, rather like the plastic protective tips on the ends of shoestrings. The telomere provides a "docking zone" or starting sequence for the replication enzymes that perform chromosome replication each time a cell divides. Telomeres also provide an inert region that prevents exposed chromosome ends from sticking to one another, which would greatly disrupt their DNA function. (Note that mtDNA has no telomeres and needs none, because it is a circular loop with no exposed ends, and instead it has two special regions where replication starts.)

Young cells, it was found, have quite long telomeres that contain perhaps 1,600 repetitions of the TTAGGG sequence. However, each time a cell divides, the replication enzyme does not replicate the telomere sequence on which it initially "parks", and so the overall telomere region is incompletely reproduced. For this reason, the telomere becomes progressively shorter with each cell division. Finally, the entire telomere is used up, the replication enzyme cannot dock, the cell can divide no further, and the unprotected chromosome ends may begin to stick together and fuse, causing DNA disruption. The affected cell malfunctions, goes senescent, or dies. When this happens to

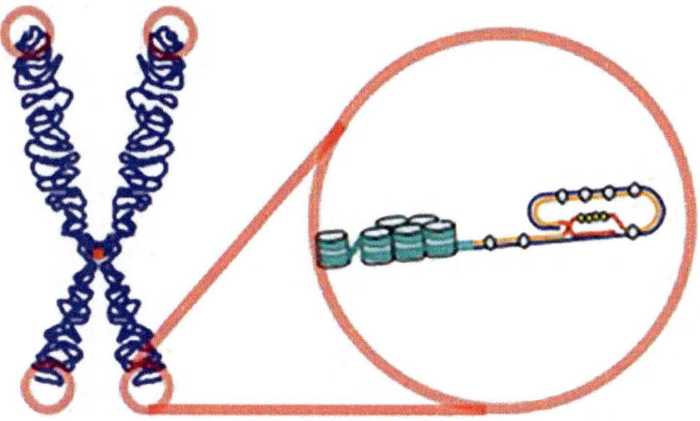

Fig. 7.1 Telomeres are repeating structures at the ends of chromosomes. *Credit* Wikimedia Commons

enough cells, the parent organism declines and shows the effects of aging [Ca14, Sa11, Sc22, Wi10].

In the 1990s, following the discovery that telomeres become progressively shorter, a widely held theory of aging was that telomere length reduction was its fundamental cause. It was heralded to the public that the ticking countdown clock for human aging had been discovered, and that a definitive remedy for human aging was just a few years away.

The enzyme *telomerase*, which has the functions of rebuilding and lengthening telomeres, became a target of interest in the longevity community. Bio-startups like the Geron Corporation promised to extend the human lifespan by rapidly developing techniques for rebuilding telomere length. However, subsequent experiments testing the telomere-exhaustion theory and its expected reversal with telomerase demonstrated that transgenetic mice engineered to have elevated levels of telomerase showed increased cancer incidence but *did not live longer*.

The implication of this work is that while telomere length is important and may be a driver of senescent cell formation, it is *not* the key to aging and its reversal. Experiments have shown that short telomeres are somewhat correlated with incipient cell senescence, and the reduction in telomere length accompanying cell division may be a roadblock for extending the human lifespan. However, it is *not* the fundamental mechanism behind human aging.

Fortunately, there are ways [Je11, Si21] in which short telomere length can be fixed when the condition is present in aging human subjects. There are currently available commercial drugs [Te25] that claim to lengthen telomeres. Alternatively, an mRNA vaccine (similar to the ones developed for COVID) could be produced that contained the RNA-coded instructions for ribosome synthesis of the catalytic protein *hTERT*, which is a protein subunit of telomerase [Sh24]. With a suitable mRNA delivery mechanism, it is likely that the arrival of this mRNA in a cell's cytoplasm should result in telomere lengthening of all chromosomes in the cell's nucleus.

References

[Ba11] D. J. Baker, et al, "learance of p16Ink4a-positive senescent cells delays ageing-associated disorders", ***Nature 479***, 232–236 (10 November 2011).

[Ba16] D. J. Baker, B. G. Childs, M. Durik, M. E. Wijers, C. J. Sieben, J. Zhong, R. A. Saltness, K. B. Jeganathan, G. C. Verzosa, A. Pezeshki, K. Khazaie, J. D. Miller, and J. M. van Deursen, "Naturally occurring p16Ink4a-positive cells shorten healthy lifespan," ***Nature 530***, 184–189 (2016).

[Ca14] D. Campa, et al, "Leukocyte telomere length in relation to pancreatic cancer risk: a prospective study," ***Cancer Epidemiol Biomarkers Prev 23***(11): 2447–2454 (2014).

[Ca25] L. de L. Camillo, et al, "A single factor for safer cellular rejuvenation", BioRXiv preprint, https://www.biorxiv.org/content/10.1101/2025.06.05.657370v1 (2025).

[Ci24] A. Cipriano, M. Moqri, S. Y. Maybury-Lewis, R. Rogers-Hammond, T. A. de Jong, A. Parker, S. Rasouli, H. R. Schöler, D. A. Sinclair, and V. Sebastiano, "Mechanisms, pathways and strategies for rejuvenation through epigenetic reprogramming," ***Nature Aging 4***, 26 (2024).

[Co05] I. M. Conboy, M. J. Conboy, A. J. Wagers, E. Girma, I. L. Weissman, T. A. Rando "Rejuvenation of aged progenitor cells by exposure to a young systemic environment," ***Nature, 433***: 760–764, (2005). https://doi.org/10.1038/nature03260. PMID: 15716955

[Co13] M. J. Conboy, I. M. Conboy, T. A. Rando, "Heterochronic parabiosis: Historical perspective and methodological considerations for studies of aging and longevity," ***Aging Cell 12***, 525–530 (2013). https://doi.org/10.1111/acel.12065. PMCID: PMC4072458.

[Cr24] S. Crespo-Garcia, F. Fournier, R. Diaz-Marin, et al., "Therapeutic targeting of cellular senescence in diabetic macular edema: preclinical and phase 1 trial results" ***Nat. Med. 30***, 443–454 (2024); https://doi.org/10.1038/s41591-024-02802-4.

[Fo18] N. Folguera-Blasco, E. Cuyàs, J. A. Menéndez, and T. Alarcón, "Epigenetic regulation of cell fate reprogramming in aging and disease: A predictive computational model," ***PLOS Computational Biology*** March 15, 2018; https://doi.org/10.1371/journal.pcbi.1006052.

[Je11] B. B. de Jesus, K. Schneeberger, E. Vera, A. Tejera, C. B. Harley, and M. A. Blasco, "The telomerase activator TA-65 elongates short telomeres and increases health span of adult/old mice without increasing cancer incidence," ***Aging Cell 10*** (4): 604–12 (2011).

[Ke21] C. Kerepesi, B. Zhang, S-G Lee, A. Trapp, and V. N. Gladyshev, "Epigenetic clocks reveal a rejuvenation event during embryogenesis followed by aging", ***Sci. Adv. 2021***; 7 (2021).

[Sa11] E. Sahin, et al, "Telomere dysfunction induces metabolic and mitochondrial compromise," ***Nature 470*** (7334): 359–365, (2011).

[Sa20] T. J. Sarkar, M. Quarta, S. Mukherjee, A. Colville, Paine, L. Doan, C. M. Tran, C. R. Chu, S. Horvath, L. S. Qi, N. Bhutani, T. A. Rando, and V. Sebastiano, "Transient non-integrative expression of nuclear reprogramming factors promotes multifaceted amelioration of aging in human cells," ***Nature Communications 11***, 1545 (2020).

[Sa21] T. D. Saccon, R. Nagpal, H. Yadav, M. B. Cavalcante, A. D. C. Nunes, A. Schneider, A. Gesing, B. Hughes, M. Yousefzadeh, T. Tchkonia, J. L. Kirkland, L. J. Niedernhofer, D. Robbins, and M. M. Masternak, "Senolytic Combination of Dasatinib and Quercetin Alleviates Intestinal Senescence and Inflammation and Modulates the Gut Microbiome in Aged Mice," ***J. Gerontol. A: Biol. Sci. Med. Sci. 76*** (11), 1895–1905 (2021).

[Sc22] C. Schneider, et al. "Association of Telomere Length With Risk of Disease and Mortality," ***JAMA Intern Med. 182*** (3) (2022).

[Sh24] H. S. Shim, et al, "TERT activation targets DNA methylation and multiple aging hallmarks," ***Cell 187*** (15) 4030–42 (2024).

[Sh25] Shift Bioscience Ltd., The Gurdon Institute, Tennis Court Road. Cambridge CB2 1QN, UK; https://www.shiftbioscience.com

[Si21] G. Singaravelu, C. B. Harley, J. M. Raffaele, Sudhakaran, A. Suram, "Double-Blind, Placebo Controlled, Randomized Clinical Trial Demonstrates Telomerase Activator TA-65 Decreases Immunosenescent CD8+CD28- T Cells in Humans," ***OBM Geriatrics 5*** (2) (2021).

[Te25] Telos Biotech, 19 Morris Ave, Brooklyn Navy Yard Bldg 128, Brooklyn, New York, 11205 USA, https://www.telosbio.com.

[Un25] UNITY Biotechnology, 285 E. Grand Ave., South San Francisco, CA 94080 USA; https://unitybiotechnology.com; Rubedo Life Sciences, 428 Oakmead Parkway, Sunnyvale, CA 94085 USA; https://www.rubedolife.com.

[Wi10] Willeit, *et al.* "Telomere length and risk of incident cancer and cancer mortality," ***JAMA 7***, 69–75 (2010).

[Zh23] Bohan Zhang, et al., "Multi-omic rejuvenation and lifespan extension upon exposure to youthful circulation," ***Nature Aging 3***, 948–964 (2023).

8

Actually Understanding and Reversing Aging

8.1 Repair and Replacement Mechanisms and Energy Demands

As we have emphasized throughout this book, there is a definite energy cost for every major cellular process in the human body, and these costs cannot be ignored. A quantity of energy (~ 0.33 eV) is released when an ATP molecule emits a phosphate molecule. Table 8.1 tabulates some of these energy costs in MU (mitochondrial units), where 1 MU = 10^8 ATP energy releases.

The mitochondrial unit (MU) defined above and used in Table 8.1 represents roughly the quantity of ATP energy that one healthy mitochondrion produces in about half a day. Clearly, in cells with many healthy mitochondria, there would be enough ATP to perform all of the above operations. However, cells with damaged and failing mitochondria will have available orders of magnitude less, leading to problems. Such energy costs cannot be ignored in considering the behavior of cells in normal operation and in reduced operation as they progressively age.

8.2 Silencing High-Demand Genes and Processes

The mechanism that drives the observed age-related epigenetic changes remains a deep biological mystery. If you ask the "experts", the usual response is "It just happens." However, there is good evidence that many nuclear DNA

J. G. Cramer, *How to Live Much Longer*, Copernicus Books,
https://doi.org/10.1007/978-3-032-17741-4_8

Table 8.1 The energy costs of various cellular processes

Process	Energy cost (MU)	Comments
Yamanaka reprogramming	~ 3–5 MU	Involves epigenetic remodeling, transcription factor activation, and metabolic shifts during OSKM-induced pluripotency.
Cell division	~ 1.2–1.5 MU	Requires ATP for DNA replication (≈50 ATP/ base pair), cytoskeletal reorganization, and membrane synthesis.
Mitochondrial replication	~ 0.8–1 MU	Includes mtDNA replication (≈1000 ATP/ genome copy) and membrane biogenesis.
Apoptosis	~ 0.5–0.7 MU	Energy-dependent process involving caspase activation (≈10^6 ATP), DNA fragmentation, and phagocytosis signaling. Note that it is somewhat more efficient than necrosis cleanup.
Necrosis	0 MU	Passive cell death uses no ATP energy. Energy costs arise only later, during the subsequent cellular cleanup.
Cleanup after necrosis	~ 0.6–0.9 MU	Phagocytosis (≈10^8 ATP), lysosomal degradation, and inflammation regulation by neighboring cells/macrophages.
Transition to senescence	~ 0.8–1.2 MU	AMPK/p53 activation, SASP initiation, and chromatin remodeling. Increased glycolysis compensates for mitochondrial dysfunction.
Senescent cell operation	~ 1.5–2 MU/day	Elevated metabolic activity for SASP secretion (cytokines, proteases, and stress responses). Less efficient energy production due to mitochondrial decline.
Normal somatic cell operation	~ 1 MU/ day	Baseline ATP usage for ion transport (Na+/K+) ATPase ≈30%, protein turnover (≈25%), and organelle maintenance.

genes are epigenetically silenced as a result of mitochondrial dysfunction [Ry24].

This leads us to an interesting hypothesis: *the progressive cellular ATP energy shortage that arises from progressively increasing mtDNA damage promotes the epigenetic silencing of genes associated with high energy consumption and the epigenetic activation of genes that perform roughly equivalent functions with lower energy consumption* [We25c]. This concept makes good evolutionary sense, for it would permit living organisms to function and compete for longer before the mitochondrial diminishment of available energy finally took its toll. The implication is that mitochondrial condition is a primary driver of epigenetic reprogramming.

8.3 Mitochondria and Sleep

Recently, it has been demonstrated from animal tests that the mitochondria in brain neurons play a key role in our sleep patterns. Researchers at Oxford [Sa25] found that when fruit flies were sleep-deprived, a specific group of neurons showed dramatic changes in their mitochondria. After a day of high energy use, the mitochondria of brain neurons have been running at top speed to produce ATP energy, and they come under stress, leading to increased ROS. Critical neurons that control the urge to sleep sense the ROS increase and respond by making us feel sleepy. In other words, sleep pressure, what makes us feel sleepy, builds up when mitochondria in key neurons of the brain are overworked and increase their ROS production.

Even more striking, they were able to dial sleepiness up or down by changing how the mitochondria were configured. When the mitochondria were forced to fission into smaller fragments, their net inner surface area increased to accommodate more Complex V sites, making more ATP, and the fruit flies slept less. When mitochondria were caused to fuse into longer structures with reduced inner surface area, the flies felt a stronger need to sleep. This suggests [Mi25a] that aging humans need more sleep because of the deteriorated condition of the mitochondria in their brains, and that if and when these are replaced with healthy mitochondria, we can expect more wide-awake subjects.

8.4 Why Do we Grow Fatter as we Age?

As we have all noticed, as we age there is a tendency to accumulate fat, particularly around the abdomen, thighs, and buttocks. Much of this can be traced to the growing inefficiency of our aging mitochondria. The nutrients provided by our stomach and intestines move to our bloodstream and go into our cells to be metabolized for energy. But as we age, there may not be enough functional mitochondria to process all of these provided nutrients. Instead of metabolizing the nutrients for ATP energy, the mitochondria-poor cell becomes a white fat cell (adipose) that stores the nutrients as lipid molecules for later metabolism. Even if you provide your body with healthy exercise and sufficient sleep, that may not matter if your mtDNA is already badly damaged, and the fat will build up by default.

Mitochondrial damage also reduces fat-burning. In the trend toward obesity, the mitochondria within white fat cells are already rather small in number, and with age, they often become damaged and less efficient, cell by cell, at burning energy. This leads to reduced fat consumption for energy and increased lipid storage. Mitochondrial dysfunction can also interfere with insulin signaling, which impairs glucose uptake in tissues like muscles and fat.

Dysfunctional mitochondria produce excessive ROS, especially in the liver. Further, damaged mitochondria can't produce enough ATP through their normal oxygen-dependent pathways and compensate by burning more glucose anaerobically through the process of *glycolysis*. Instead of producing ~ 32 ATP molecules per glucose molecule consumed, they make only ~ 2 ATP per glucose and generate lactate waste products. This stress contributes to insulin resistance, which further disrupts glucose metabolism and promotes more fat accumulation. Obesity itself floods cells with nutrients, and the mitochondria adapt by becoming more efficient at storing energy as lipid fat and less efficient at burning it to produce ATP. This functional shift favors fat buildup.

It's basically a feedback loop: mitochondrial dysfunction causes metabolic alterations that favor fat storage over ATP production. The resulting obesity and nutrient overload in turn induce mitochondrial dysfunction, leading to more obesity, and around the loop. To break this cycle, it would be necessary to arrange for an influx of healthy mitochondria with unmutated mtDNA.

8.5 Mitochondrial Damage

Mitochondrial damage and its exponential rise is discussed in more detail in Appendix B below. Briefly, mitochondrial can be damaged by many different agents: chemotherapy, stress, smoking, ionizing radiation, local radioactivity, some common drugs, anesthetics during surgery, physical injury, and worst of all, simple old age. The mtDNA is the body's weak link, much more fragile and more easily damaged than the other components of cells. Further, research indicates that mtDNA decline increases *exponentially* as we age, (see Appendix B) inexorable and unstoppable, outrunning intrinsically linear cellular repair and replacement processes.

Our current medical system has no significant recognition of any of these problems, because at present, mitochondrial decline is medically invisible. There are no standard medical tests for mitochondrial dysfunction or mtDNA damage. There are a plethora of remedies involving diets, supplements, or vitamins that claim to help mitochondria, but with no tests, they are all just shots in the dark that cannot repair mtDNA damage. True mitochondrial testing is available in research, but it has not been standardized, and all such tests are presently much too expensive for general use.

8.6 Senescent Cells and Senolytics: Apoptosis Versus Necrosis

With age, many cells become senescent due to damage and are "retired in place", blocked from dividing, perhaps continuing some aspects of their prior functions, using around twice as much energy as before going senescent, and secreting toxic SASP that causes inflammation and damages neighboring cells. Senolytic drugs, as discussed in Chap. 8 Sect. 8.1 above, cause at least some senescent cells to go into apoptosis, a "clean" cell death and removal with little debris. However, as we see from Table 8.1, apoptosis requires the energy supplied by about 600,000,000 ATP molecules.

For older humans, within a given senescent cell that quantity of energy may not be available. In that case, there is an alternative with no immediate energy cost: the cell goes into *necrosis*, a "messy" cell death that releases toxins and debris and requires subsequent energy-expensive cleanup.

For about the last four years, I have done three-day D + Q + F senolytic sessions (described in Chap. 8) every 3–4 months to support my own longevity. These sessions have typically caused me to experience the short-lived symptoms of a mild headache, low fever, stomach cramps, daytime sleepiness, and a general feeling of unwellness, rather like coming down with the flu. These symptoms characteristically lasted only about a day and were followed by weeks of feeling very well. I had always assumed that the short-lived negative side effects were just an inevitable part of the senolytic damaged-cell-removal process.

However, recently I did 13 hours of 1.5 atmosphere HBOT sessions over three days, a treatment that is expected to at least temporarily but significantly increase the body's supply of ATP energy. The week following this treatment, I again did three days of D + Q + F senolytic sessions. Interestingly, I felt very well immediately and did *not* experience any of the usual negative flu-like side effects. This may all be just the placebo effect in action, but I think that this time my body had enough ATP to do a complete job of apoptosis of the targeted senescent cells, and that I avoided having any of them going into necrosis and put toxins and cell fragments in my bloodstream instead. This, of course, isn't proof of anything, but it reinforces my conviction that loss of available ATP energy is the key to understanding human aging.

8.7 High Volume Mitochondrial Transplantation

If the primary cause of the effects of human aging is the declining condition of the subject's mitochondria leading to a shortage of ATP energy, a clear solution for such problems would be to replace a sizable fraction of the failing mitochondria with healthy ones. It is presently not clear how high that fraction needs to be to achieve significant age reversal. Mitochondrial multiplying effects from stimulated mitogenesis are possible, which would reduce the needed transplant size. My educated guess, taking into account a possible boost from stimulated mitogenesis, is that the target should be a replacement level of 10% or more. For particular organs [Ch25], it has already been demonstrated that the introduction of new mitochondria can heal local damage. However, general large-volume replacement of damaged mitochondria in human subjects has never been tried. That intervention is called *high-volume mitochondrial transplantation*.

It is estimated that mitochondria, although individually very small, account for about 10% of a human's total body weight because of their great numbers in human cells. Therefore, in a very old human subject that might need replacement of roughly 10% or more of their failing mitochondria, so the quantity of fresh mitochondria needed for the transplant would be measured in *kilograms*. Obtaining and transplanting such a large volume (liters) of mitochondria is not impossible, but it is technically challenging. In particular, it would likely require industrial-scale production of new mitochondria. In principle, this could be accomplished by multiplying stem cells that contain healthy mitochondria in a large bioreactor and then extracting and transplanting their mitochondrial content.

8.7.1 Mitochondrial Source and Quality Control

To begin that process, from what source should the starter stem cells be obtained? The seed cells could be obtained from blood, bone marrow, muscle tissue, or liposuctioned fat cells. Basically, there are four choices here for the source: (1) obtain the stem cells from the subject himself. This has the problem that if the aged subject already has a large fraction of damaged mitochondria, the same is likely to be true of most extracted tissue; (2) obtain the stem cells from a younger maternal-line relative of the subject, insuring that the cells have the same mtDNA haplogroup; (3) obtain the stem cells from an unrelated donor that had been matched to have the same or similar mitochondrial haplogroup as that of the subject; or (4) obtain the stem cells from a random donor without haplogroup matching. The last option might be part of a mass-produced mitochondrial bank prepared for general emergency treatments and for subjects unable to pay for a more specialized and targeted mitochondrial transplant.

Before use in a bioreactor replication, it may be necessary to select the starter stem cells for replication of only those containing mitochondria with relatively undamaged mtDNA and healthy respiration. We note that proprietary procedures already exist for performing such selection, for assuring vigorous stem cell replication, and for adding perfect mtDNA copies after stem cell replication, but we will not describe them here.

8.7.2 Mass Replication in Bioreactors

As stated above, a presently available technique for large-scale replication of cells is through the use of a bioreactor. Bioreactors are already widely used in producing stem cells for use in stem cell therapy and for stem-cell-derived exosome therapy, as previously discussed in Chap. 7 (Fig. 8.1).

Fig. 8.1 A large volume bioreactor. *Credit* U. S. Dept. of Energy, Public Domain

A typical bioreactor maintains precise internal temperature, pH, oxygen level, and nutrient supply to mimic ideal natural conditions. Sensors continuously track the cell growth, adjusting conditions in real time to optimize quality and yield. The cells being multiplied receive gentle mixing to ensure uniform distribution of nutrients and oxygen while minimizing shear forces that could damage the delicate cells. Bioreactors, as described, are scalable to very large volumes for industrial-scale production. There are also emerging bioreactor techniques involving permeable hollow fibers that may prove to be more efficient.

The eventual goal of this effort would be to produce bioreactor-grown mitochondria at an industrial scale, creating a medical "tool kit" of many different haplotypes of mitochondria that could be used by clinicians to treat emergencies and regenerate various failing organs, produced in quantities sufficient to supplement every human being over the age of perhaps 55 and to reduce in severity or eliminate a wide array of genetic and other diseases.

8.7.3 Mitochondrial Extraction and Mitlet Encapsulation

After the target replicated volume of stem cells is achieved, they are removed from the bioreactor, and the mitochondria they contain are extracted and isolated. The mitochondria then need to be encapsulated within protective *mitlets* to avoid interaction with the recipient's immune system. Mitlets containing mitochondria are a "natural fountain of youth" within the body, used to balance and preserve healthy cellular energetics. They are an extraordinary evolutionary adaptation to increase the survival and longevity of species like humans. They can be isolated, stored, transplanted, grown, and manipulated, presenting broad potential for disease-modifying treatments.

8.7.4 Mitochondrial Delivery: General Versus Targeted?

There are many different techniques for transplanting mitochondria to tissues in the body. In animal studies, cells in need have been found to readily take in transplanted mitochondria and put them to use. Since creating a mitochondrion requires considerable local energy (see Table 8.1), this supply from outside using external energy conveys a significant evolutionary and survival advantage.

Mitrix Bio has tested such transplantation in aged mice and has observed excellent and predictable results in the immune system, retina, brain, and skin. For example, young mitochondria transferred into the bloodstream of older mice have been shown to reliably "reverse the age" of the immune system, with dramatic improvements in the survival of mice confronted with life-threatening viral or bacterial infections.

8.7.5 Mitochondrial Cycle Theory

Not only do mitochondria transfer naturally throughout the body, but this process seems to be "intelligently" controlled. The Mitochondrial Cycle idea, which has been proposed and promoted by Mitrix Bio, is that the human body has large reservoirs of mitochondria with young mtDNA in the bone marrow or other areas, which are slowly replicated and transported by the bloodstream to peripheral tissues over time to replace damaged mitochondria. This supplemental flow of mitochondria [Bo23a] is one of evolution's tools for producing long-lived species like humans. This hypothesis seems to be supported by the observation, as described in Chap. 5 Sect. 5.10, that the mtDNA of the white blood cells of a 90 year-old human subject (the author) are much less damaged than the mtDNA of the subject's kidney cells.

8.8 Follow-Ups to Mitochondrial Transplantation

Since there are as yet no human subjects who have received large-volume mitochondrial transplantation, it is not clear whether that intervention is the *only one needed for significant age reversal*, or if other follow-up interventions will also be required. The subject's severe shortage of cellular ATP energy will presumably be decisively eliminated, but it is not known if hypothetical repair and youth-restoration effects will follow "automatically" from the elimination of the severe energy shortage. Some evidence from the effects of parabiosis on old mice suggests that this could happen, but real results on human subjects need to be obtained and evaluated.

8.8.1 What Is Automatic? What Requires Intervention?

In particular, there are several unanswered questions about the results of a large-volume mitochondrial transplantation in a very old subject: (1) Does the transplant recipient still have significant quantities of senescent cells in various organs, causing inflammation and releasing SASP? (2) Does the recipient still have "old" epigenetic programming reflecting his calendar age? (3) Does the recipient have missing cells that have not been replaced when senescent cells were cleared? and (4) Does the recipient still have dangerously over-short telomeres reflecting his calendar age? Each of these issues can be identified and treated with a specific intervention as a follow-up, if they are found to be present.

8.8.2 Senolytics: Use the New Energy to Clear the Junk?

As discussed in Chap. 7, Sect. 7.1.1 above, senescent cells may be cleared by drugs or plasmids that trigger them to go into apoptosis. Apoptosis is a somewhat energy-demanding process, possibly making it less desirable in treating the very old, even if they are suffering from inflammation and have acquired a significant fraction of cells that are senescent.

However, following large-volume mitochondrial transplantation, ATP energy should be abundant. Therefore, if tests show that the subject still has a significant population of senescent cells, effective senolytics would be very appropriate. In my opinion, the best senolytic procedure is the one from Oisin Bio [Oi25], which has developed a plasmid that triggers apoptosis if the cell is attempting to express p16, a marker protein for senescent cells that suppresses cell division. This intervention has not yet received FDA approval for humans, but it has been found to work very well on aging mice. The use of D + Q + F senolytics described in Chap. 7, Sect. 7.1.1 above is a viable alternative.

8.8.3 Epigenetic Reprogramming: "Younger" Protein Expression?

As discussed in Chap. 3 Sect. 3.1, the nuclear DNA is "marked" by the attachment at particular spots of a methyl CH_3 group to the nucleobase cytosine, which has the effect of silencing the related gene. This is called *epigenetic programming*, and it determines which of the many possible proteins encoded in the nuclear genome are being produced and which are not.

Following large-volume mitochondrial transplantation, the state of this epigenetic programming will need to be determined. If the increased abundance of ATP energy has not automatically restored a "young" profile to the epigenome, then an epigenetic reprogramming intervention should be implemented to produce one. As discussed in Chap. 3 Sect. 3.1, there is a new epigenetic reprogramming protein developed by Shift Biosciences in the UK [Sh25] called SB000, which, if it proves to be without problems, should be able to achieve whole body epigenetic reprogramming to a younger state [Ca25]. If SB000 has unforeseen difficulties, the better-known Yamanaka-based techniques can be used to reach a similar goal.

8.8.4 Stem Cell Therapy: Replace Cleared Cells with New Ones?

As discussed in Chap. 6 Sect. 6.9, there is a well known intervention that involves obtaining a sample of stem cells from either the recipient or a young donor, multiplying them in a bioreactor as described above, and re-injecting them into the recipient, either at selected body locations in need or restoration (eye, knee, liver, brain, ...) or into the bloodstream for general distribution.

In the case of stem cell therapy as a follow-up to the mitochondrial transplant, successful whole-body senolytics may result in many cell "vacancies" in organs in which senescent cells have been cleared but not replaced. In such a situation, if the newly available ATP energy does not stimulate the cell-depleted organs to generate new cells, making bioreactor-derived stem cells available to implement replacement might prove to be an effective treatment.

8.8.5 Is Telomere Lengthening Necessary?

As discussed in Chap. 5, Sect. 5.2, with each cell division the telomeres of each of the nuclear DNA chromosomes get progressively shorter and shorter. Following a hypothetical large-volume mitochondrial transplantation, it is not clear if the newly available supply of ATP energy would automatically stimulate cellular repair processes to implement general telomere lengthening, as some evidence from exosome therapy might suggest. If not, a whole-body intervention to produce longer telomeres in aging subjects may be required. This might be done in a variety of ways by activating the enzyme telomerase to restore short telomeres to a reasonable length. Fortunately, the cellular energy cost for such telomere extension for one cell nucleus is relatively modest, requiring at maximum the energy supplied by about 2,000 ATP molecules. Even for very old subjects, this is a very low energy requirement that should be easy to supply.

Telos Biotech [Te25] and Telomir Pharmaceuticals [Te25a] are both commercial firms making products that claim to restore telomere length by activating telomerase. An intervention using such drugs could, in principle, produce whole-body lengthening of telomeres, if that was required.

An untested alternative might be to synthesize a form of messenger RNA (similar to the mRNA COVID vaccines) that has the coding for the protein hTERT, the presence of which in cells triggers the production of telomerase, which lengthens telomeres.

References

[Bo23a] N. Borcherding and J. R. Brestoff, "The power and potential of mitochondria transfer," ***Nature 623***, 283-291 (2023); https://doi.org/10.1038/s41586-023-06537-z.

[Ca25] L. de L. Camillo, et al, "A single factor for safer cellular rejuvenation", BioRXiv preprint, https://www.biorxiv.org/content/10.1101/2025.06.05.657370v1 (2025).

[Ch25] Xuri Chen, Yunting Zhou, Wenyu Yao, Chenlu Gao, Zhuomin Sha, Junzhi Yi, Jiasheng Wang, Xindi Liu, Chenjie Dai, Yi Zhang, Zhonglin Wu, Xudong Yao, Jing Zhou, Hua Liu, Yishan Chen & Hongwei Ouyang, "Organelle-tuning condition robustly fabricates energetic mitochondria for cartilage regeneration," ***Bone Research 13***, 37 (2025).

[Mi25a] F. A. Mir, A. R. S. Lark, and C. J. Nehs, "Unraveling the interplay between sleep, redox metabolism, and aging: implications for brain health and longevity," ***Frontiers of Aging 6***, 1605070; https://doi.org/10.3389/fragi.2025.1605070.

[Oi25] Oisin Biotechnologies, 701 Fifth Ave, Suite 4200, Seattle, WA 98104; https://www.oisinbio.com.

[Ry24] Keun Woo Ryu, Tak Shun Fung, Daphne C. Baker, Michelle Saoi, Jinsung Park, Christopher A. Febres-Aldana, Rania G. Aly, Ruobing Cui, Anurag Sharma, Yi Fu, Olivia L. Jones, Xin Cai, H. Amalia Pasolli, Justin R. Cross, Charles M. Rudin & Craig B. Thompson, "Cellular ATP demand creates metabolically distinct subpopulations of mitochondria," ***Nature 635***, 746–754 (2024).

[Sa25] R. Sarnataro, C. D. Velasco, N. Monaco, A. Kempf, and G. Miesenböck, "Mitochondrial origins of the pressure to sleep," ***Nature*** 16 July(2025); https://doi.org/10.1038/s41586-025-09261-y.

[Sh25] Shift Bioscience Ltd., The Gurdon Institute, Tennis Court Road. Cambridge CB2 1QN, UK; https://www.shiftbioscience.com.

[Te25] Telos Biotech, 19 Morris Ave, Brooklyn Navy Yard Bldg 128, Brooklyn, New York, 11205 USA, https://www.telosbio.com.

[Te25a] Telomir Pharmaceuticals, Inc., 100 SE 2nd Street, Suite 2000 #1009, Miami, FL 33131, United States; Tel: 786 396 6723; https://telomirpharma.com.

[We25c] V. Weissig, "Mitochondrial dysfunction as the "mother" of all hallmarks of aging," ***Journal of Mitochondria, Plastids and Endosymbiosis, 3:1,*** 2560191, (2025); https://doi.org/10.1080/28347056.2025.2560191.

9

Engineering Methuselah?

In this book we have been advocating the view that the primary source of human aging comes from accumulated damage to the 37 genes encoded in the small and vulnerable ring of mtDNA. Only 13 of these mtDNA genes actually code for needed mitochondrial component proteins. The other 24 genes are present to support the ribosomal production within the mitochondrion of those 13 key proteins. This arrangement presents an interesting question: *Since evolutionary pressure has already moved almost all of the genes coding for mitochondrial components to the safer domain of nuclear DNA, what if we used genetic engineering to relocate the remaining 13 genes there?* Would this create a new race of human "Methuselahs", a new subspecies of humanity that would age very slowly and would live for perhaps thousands of years without the need for strong longevity interventions? This idea was first proposed decades ago by prominent age-reversal advocate Aubrey de Grey. Aspects of it have since been investigated by his SENS Foundation [Gr99].

Implementing the proposal to transfer the 13 protein-coding genes from human mitochondrial DNA to the nuclear DNA would create profound challenges and new opportunities in genetic engineering. While the cell nucleus already contains thousands of mitochondrial component genes, integrating these remaining 13 genes, which are crucial for the mitochondrion's core function of ATP production, raises significant questions about transport, viability, adaptation, and long-term benefits, including the intriguing possibility that it would eliminate many mitochondrial genetic diseases and greatly extend the human lifespan.

J. G. Cramer, *How to Live Much Longer*, Copernicus Books,
https://doi.org/10.1007/978-3-032-17741-4_9

9.1 Ethical Issues

In our present worldwide practice of medicine, there are strong ethical prohibitions against implementing *any* inheritable genetic engineering changes in the human species, and these prohibitions would certainly be breached by any such changes and rearrangements of the human genome [Pa24]. It is also not clear whether humans with the contemplated genetic changes might be isolated as pariahs that could not breed with the normal human population. Nevertheless, as an intellectual exercise, it is interesting to consider the possibility and its consequences. In some parts of the world, human inheritable genetic engineering experiments of questionable ethics have already been performed, and undoubtedly there will be more.

9.2 Moving the 13 Vulnerable Genes out of mtDNA

As stated above, human mtDNA contains a small but vital set of 13 genes that directly code for sub-units of the ATP energy-generating machinery of the cell. These genes are considered "vulnerable" because mtDNA operates under more hostile conditions than those of nuclear DNA (see Appendix 2). The mitochondrion has somewhat error-prone DNA replication and lacks the robust DNA repair mechanisms found in the cell nucleus, leading to a higher mutation rate by around a factor of 20 or more. The mtDNA is exposed to ROS and other damage, is susceptible to replication errors, and is provided with minimal repair machinery. Moving these 13 genes to the cell nucleus would hypothetically protect them from these stress elements, offering the benefit of the nucleus's better protection and more comprehensive DNA repair systems.

The word "move" in the proposal may be deceptive. The actual proposed strategy seems to be to insert suitably modified copies of the 13 protein-coding genes in the nuclear DNA, while leaving the old redundant genes still in place in the mitochondrion. Then, if the mtDNA genes suffered damage, the protein components needed for ATP production would still be supplied by the nuclear DNA. Thus, if there are other important functions performed by the mtDNA, particularly in the somewhat mysterious D-loop control region, these would still be in place.

9.3 Coding Changes

A significant hurdle in transferring mitochondrial genes to the nucleus lies in the three-letter genetic code itself. The mitochondrial genetic code differs slightly from the universal genetic code used in the DNA of the cell nucleus and elsewhere. Certain codons that specify a particular amino acid or a stop signal in the nuclear DNA have different meanings in mtDNA. Therefore, a direct literal transfer of the mitochondrial genes to the nucleus would result in incorrect protein synthesis from misinterpreted codons.

To ensure proper function, these 13 genes would need to be re-coded, with incompatible mtDNA codons modified ("nuclearized") to match the nuclear genetic code, allowing the cell's ribosome translation machinery to produce the correct protein sequences. This might be done by synthetically sequencing the DNA to be inserted in the cell nucleus from a digital script that includes the needed coding changes, rather than by attempting to re-code and relocate segments of the existing mtDNA.

9.4 Epigenetic Regulation

Nuclear genes are subject to complex layers of epigenetic regulation, including CpG methylation, acetylation, histone modification, and non-coding RNA interactions, which tightly control nuclear gene expression. The mtDNA genes, in contrast, have their own simpler regulatory mechanisms and are never "silenced" by epigenetics. If the 13 mtDNA genes were moved to the nucleus, they would need to be integrated into the complex epigenetic regulatory network there. This would involve ensuring that they were equipped with appropriate nuclear promoters, enhancers, and other regulatory elements to achieve correct levels of expression, at the right time, and in the right tissues, mirroring their essential roles in ATP production within the mitochondrion. Without proper nuclear regulatory control, the proteins might be overproduced or underproduced, leading to cellular dysfunction from improperly controlled mitochondrial processes.

9.5 Import Efficiency and Compatibility

All 13 of the proteins whose genes are to be moved have the characteristic of being very *hydrophobic*, meaning that they tend to avoid water. This would make it very difficult to transport the ribosome-produced proteins from the water-based cell cytoplasm, where they had been assembled, into the mitochondrial interior. That is perhaps the reason why long ago these genes had not been relocated to the cell nucleus along with the thousand or so other mitochondrial genes.

In particular, to avoid as much contact with water as possible, the hydrophobic proteins released into the cytoplasm would probably fold themselves into small balls of minimal surface area. Upon arrival at the location in the mitochondrial interior where they were to become part of the ATP production machinery, they would have to be unrolled and refolded, a process that would not happen automatically. There are perhaps ways of dealing with this issue, but all of them would be difficult.

One strategy might be to add extra amino acids to all 13 of the hydrophobic proteins, so that, while still performing the same functions with the same efficiency in mitochondrial ATP production, they are more *hydrophilic* and can move freely in the water-based cellular cytoplasm. Such modifications would have to strike a delicate balance between hydrophilia and functionality in ATP production, and would require much work in simulation and testing on cell cultures to determine if the scheme was even feasible.

Even if these 13 genes were successfully relocated to the nucleus and properly regulated, their protein products would still need to reach their functional destinations within the mitochondria. This would require the addition of a mitochondrial targeting sequence (a positively charged peptide about 15–70 amino acids long that facilitates transport to the mitochondria) to the beginning of each protein. This sequence would act as a "zip code," directing the newly synthesized proteins from the cytoplasm to the mitochondrion.

The efficiency of this import process would be critical; the cell's existing machinery might struggle to import all 13 essential proteins as fast as was needed. Furthermore, there is the issue of "mito-nuclear compatibility." The nuclear-encoded and mitochondrial-encoded components of mitochondrial complexes have co-evolved over billions of years, leading to highly optimized interactions. Disrupting this delicate balance by changing the origin of essential components might lead to functional mismatches and reduced mitochondrial efficiency.

Retaining the 13 genes in the mitochondrion allows for tight, local regulation of the ATP production machinery in response to changes in mitochondrial conditions (like the redox state or inter-membrane potential). Also, rapid response to damage or metabolic state may require that some of the 13 proteins be assembled locally.

9.6 Assembly Issues

Many of the crucial protein complexes within the mitochondria, particularly those five involving ATP production, are composed of subunits supplied by both the nuclear DNA and mtDNA genomes. For example, some subunits of Complexes I, III, IV, and V are encoded by mtDNA, while others are nuclear-encoded. If the mtDNA encoded subunits were moved to the nucleus, they would be synthesized in the cytoplasm alongside their nuclear-encoded counterparts. Ensuring correct and efficient assembly of these multi-subunit complexes, where all components are now manufactured outside the mitochondrion, might present a significant challenge. Proper folding, amino acid delivery, and timely assembly within the mitochondrial matrix would be essential for complex functionality.

9.7 Single Gene Move Tests

There has already been one successful test of inserting a human mitochondrial gene into the cell nucleus. The mitochondrial genetic disease LHON (Leber Hereditary Optic Neuropathy) can result from a point mutation at any of three different mtDNA locations (3,460, 11,778, or 14,484). In one of these damaging mutations, blindness is caused by a G > A mutation at location 11,778 in the ND4 gene. It has been treated by implanting a new ND4 gene into the nuclear DNA. This was done by synthesizing a version of the ND4 gene plus a mitochondrial targeting sequence, with coding changes appropriate for the cellular ribosome code. An engineered AAV virus was then injected directly into one eyeball to implant a small DNA ring that included this ND4 gene sequence within the cellular nuclei. Of five legally blind participants receiving this treatment in the initial trial, two showed measurable improvements in vision, and none reported negative effects.

Experimental work has also shown partial success in relocating the ATP6, COX1, COX2, and COX3 genes. However, the COX2 relocation proved particularly difficult, due to problems with proper mitochondrial import and integration of the cytoplasm-produced protein.

There have also been some experiments with yeast cells, which have very different mitochondria from mammals. Researchers have successfully relocated ATP9 and COX2 mitochondrial genes to the nucleus in yeast, but these have required extensive recoding and chaperone support. Given the complexity and the potential risks associated with such a fundamental genetic alteration, extensive testing using animal models would be necessary before considering any more complicated human applications.

9.8 Genetic Inheritance from both Parents

If all of the 13 mtDNA genes could be relocated to the nuclear DNA, there would be a significant evolutionary implications for genetic inheritance. Mitochondrial DNA is essentially always inherited from the mother through the winnowing bottleneck process described in Appendix C below. This means that genetic mutations in the mother's mtDNA can be amplified and passed down to all of her children. Her daughters can then, in turn, pass the damaged mitochondria on to the next generation, and so on. By transferring the 13 mtDNA genes to the nuclear genome, they would become subject to the usual two-parent inheritance, i.e., they would be inherited from *both* mother and father, as are all of the other genes located in the nuclear DNA. This could have profound long-term implications for genetic disease transmission, genetic diversity, and the ultimate path taken by future human evolution.

9.9 Longevity Benefits

The hypothesis that relocating these 13 vulnerable mtDNA genes could lead to longevity benefits is based on the recent realization that mitochondrial dysfunction is a key contributor to human aging and to age-related degenerative diseases. By moving the 13 genes to the more protected and robust environment of the nucleus, the rate of damaging mutations for these essential genes should be reduced by a factor of about 20. In terms of the effect of this on human longevity, if accumulated mtDNA genetic damage was the only factor that determined maximum lifespan (there are likely to be others), these genetically modified Methuselahs should live about 20 times longer, or for around 2,000 years without requiring any further life-extending interventions.

The probability of children incurring devastating mitochondrial genetic diseases would be reduced by more than a factor of 20. Less mitochondrial damage could translate to long-term improved and more stable ATP energy production, reduced oxidative stress, and a slowing of age-related mental and physical decline. These prospects underpin the concept of "Engineering Methuselah," aiming to enhance cellular resilience and extend healthy lifespan by better protecting and repairing crucial mitochondrial functions.

However, even if all of the difficulties discussed above could be overcome, there could still be other problems. Perhaps there is a good reason why evolution has not already moved the vulnerable 13 protein-coding mtDNA genes to the better-protected nuclear DNA. Within each mitochondrion in a cell, there are multiple mtDNA copies, which are distributed and actively translated over many locations within the organelle. Presumably, they are distributed in this way for a purpose, perhaps to perform useful local functions like assembling RNA "on the spot" because it is needed there for efficient, ongoing ATP production, and perhaps also to perform other functions like intracellular communication. If the 13 mtDNA genes are moved to the more remote location of the nuclear DNA, that could result in a significant decrease in the rate and efficiency of ATP production and restoration in case of damage. Animal testing would be needed to address these concerns.

9.10 Should it Be Tried?

While the theoretical framework for moving these 13 mtDNA genes to the nucleus exists and is supported by the evolutionary history of gene transfer from mitochondria to the nucleus, the ethical challenges and practical implementation issues present immense problems for genetic engineering. Overcoming the hurdles of genetic code differences, regulatory integration, protein targeting, complex assembly, hydrophobic protein transport, and overall cellular compatibility would require sophisticated genetic engineering, animal testing that could exceed current capabilities, and extensive research and analysis.

Further, while it might benefit the children produced by this initiative, it offers little value to the current version of humanity, those of us now coping with the problems of age and decline. Clearly, this genetic engineering initiative is "not yet ready for prime time", except as an advanced experimental treatment for humans with mitochondrial genetic diseases.

(*Note:* A version of this chapter appeared as a "The Alternate View" column by the author in the November–December-2025 issue of ***Analog Science Fiction and Fact Magazine***.)

References

[Gr99] Aubrey D. N. J. de Grey, ***The Mitochondrial Free Radical Theory of Aging***, R. G. Landes Co., (1999); ISBN-13: 978-1570595646.

[Pa24] S.J. Park, Y.Y. Kim, J. Y. Han, et al, "Advancements in Human Embryonic Stem Cell Research: Clinical Applications and Ethical Issues," **Tissue Eng Regen Med 21**, 379–394 (2024).

10

The Coming Mitochondrial Revolution

10.1 Mitochondrial Transplants, Genetic Diseases, and Help in the Emergency Room

In the near future, the introduction of mitochondrial transplant and mitochondrial enhancement technologies will likely come from medical necessity rather than from anti-aging interventions. The very early recipients will probably be children and young adults who are afflicted with mitochondrial genetic diseases like MELAS and MERRF (disorders of the nervous system and muscles), LHON (optic neuropathy), and NARP (neuropathy, ataxia, and *retinitis pigmentosa.*) These mitochondrial diseases afflict more than 3,000 children born each year in the USA. They occur by random chance with low probability, when one of a mother's mitochondria with mutated mtDNA makes it past the mitochondrial bottleneck (see Chap. 3, Sect. 3.2 above and Appendix C below) and is vastly multiplied in the developing embryo. Prompt large-volume transplantation [Ad22] of healthy mitochondria should provide a "miracle remedy" for all of these mitochondrial genetic diseases, and the implementation would enrich and extend the lives of many thousands every year.

Many of the most common hospital emergency room crises (heart attacks, strokes, lung and kidney failure, severe wounds, sepsis, ...) could rapidly and effectively be treated if a supply of generic mitochondria had been previously stored and was available for rapid injection or infusion. Such prepared mitochondria could be quite feasibly stored for weeks or months, so that they were ready on demand to re-energize failing cells and to boost the body's immune and repair systems. Further, mitochondrial dysfunction underlies numerous

J. G. Cramer, *How to Live Much Longer*, Copernicus Books, https://doi.org/10.1007/978-3-032-17741-4_10

aging-related diseases, from cardiovascular deterioration to neurodegeneration, and infusion of generic mitochondria could provide some level of immediate restoration. Hospital Emergency Departments would likely become the first proving grounds for using this new technology in emergency settings. Physicians could treat patients with demanding needs that could be immediately addressed through mitochondrial transplantation, using the newly available out-of-the-cooler generic mitochondria.

The medical infrastructure needed to support this innovation would require a significant transformation. Current emergency protocols commonly assume irreversible damage. With the new mitochondrial restoration capabilities, the treatment paradigms would need to shift toward aggressive intervention rather than triage and palliative care. Hospitals might need specialized mitochondrial storage facilities, rapid haplotype matching systems, and trained personnel capable of performing infusions and targeted injections under both routine and emergency conditions. The mitochondrial resource might initially be scarce, and the implications for ethical use might be immediate and profound, as physicians would have to decide which patients should receive life-extending treatments and which should default to traditional treatments or palliative care.

10.2 The Rise of the Big Bioreactor: Cranking out the Mitochondria

Where would the needed mitochondria come from? There are many potential sources of mitochondria for transplantation (blood platelets, bone marrow, fat cells, cord blood, birth placenta, ...), but most of these options, particularly from older donors, have two limiting issues: the condition of mtDNA and the available volume of mitochondria that can be obtained from a particular source.

As discussed above, transplantation to enhance longevity would ideally use *kilograms* of haplotype-matched mitochondria containing perfect, completely unmutated and undamaged mtDNA. There are several strategies for approaching this ideal by addressing the mtDNA quality issue. It is important to realize that about 10% of human body weight is attributable to its mitochondria, so kilograms of mitochondria might be required in a given infusion to significantly boost the healthy fraction. For the large volume therefore needed, there appears to be only one major option on the immediate technical horizon: constructing *very large bioreactor* facilities that can amplify mitochondria indirectly by driving repeated cell division of selected healthy stem cells (which contain the needed mitochondria.)

Industrial-scale mitochondrial production would become essential as demand exploded beyond natural donor supply capabilities. Massive bioreactor facilities would emerge as the pharmaceutical industry's new profit frontier, requiring unprecedented biotechnological infrastructure. These facilities would operate more like semiconductor fabrication complexes than like traditional drug manufacturing plants, maintaining controlled, sterile, and carefully monitored environments capable of producing many kilograms of functional mitochondria-containing cells daily.

Quality control, testing, and sequencing would become paramount because any propagating mutation or contamination that slipped into the final product could have severe consequences. These facilities would need intensive real-time monitoring systems, automated quality testing, and a fast distribution system capable of maintaining the viability of the product during transport. The energy requirements alone might be substantial, because maintaining optimal cellular growth conditions at an industrial scale would likely demand a significant and reliable power infrastructure. Storage and transport issues might also determine whether centralized mitochondrial production facilities were feasible, or whether the bioreactors would need to be widely distributed to many population centers for fast transport and delivery.

There is also the issue of the haplotype matching requirements, which are likely to be more demanding than the more familiar blood-type matching used by blood banks. Key questions are: (1) does each mitochondrial transplantation recipient need his own unique bioreactor, producing mitochondria perfectly matched to his mtDNA sequence, or (2) should there be many large bioreactors (perhaps 10–30), each producing mitochondria for a major mitochondrial haplogroup, or (3) could there be *only one* huge bioreactor producing generic "O-type" mitochondria for all recipients? At present, the resolution of this haplogroup trichotomy is unclear, and many future implementations and economic decisions will depend on it.

10.3 O-Type Mitochondria, Haplogroup Match, or Use-Your-Own?

The fundamental question of mitochondrial compatibility would shape the entire industry structure. A universal "O-type" mitochondrion, selected or perhaps bio-engineered for broad compatibility, would enable mass production and global distribution. However, that choice might sacrifice optimal performance or create downstream problems for poorly matched individual recipients. This approach mirrors that of universal blood donors, but it has far greater complexity, given that mtDNA variations across the human population are conventionally divided into 200 or more different mitochondrial haplogroups and subgroups [Ph25, Mi25b].

Universal specific matching of the main haplogroup would require maintaining diverse production lines corresponding to many of the 42 major human mitochondrial lineages. However, about 43% of US and Canadian residents and 45% of the residents of Western Europe and Australia have the same primary mitochondrial haplogroup H, so setting up mass production only for haplogroup H would satisfy almost half of the immediate haplogroup matching needs in these countries, and haplogroup H, with a 50–50 chance of a primary match, might become the standard the generic mitochondria used in hospital emergency rooms. Several of the other mitochondrial haplogroups (U, J, T, K, ...) are found in the above geographic locations at about the 10% level, which might warrant mass-producing these also. Nations in other geographic locations having concentrations of alternative haplogroups might need to provide their populace with the locally dominant haplogroup. However, it would still be likely that many "orphaned" individuals with unpopular haplogroups would have to make do without even top-level haplogroup matching, with presently unknown consequences.

My own mitochondrial haplogroup is H2a1g1, which came from my maternal-line Norwegian great-grandmother Annie Jensen Kinzbach (~ 1843–1893). Annie and I are separated by about 53 inherited mtDNA mutations from the original "mitochondrial Eve" and by 5 inherited mutations from the top-level H haplogroup. Therefore, at least at the first level, I might be in the best-supported main group in the USA, but I would still have 5 genetic mismatches. It is unclear just how detailed the haplogroup matching needs to be to avoid mismatch issues, but just matching at the main haplogroup level would certainly be a good start.

The "billionaire's alternative" would be to extract, amplify, and reintroduce each longevity extension subject's own unique mitochondria, thereby having the subject's exact mtDNA sequence (with perhaps a few random mutations, depending on how the cells for amplification were selected). This would offer essentially perfect compatibility. However, for each such individual, it would probably require the exclusive use of one dedicated bioreactor and associated support equipment and staff for a year or more, at high complexity and cost (~ $ one million or more). This would certainly limit such access to only the extremely wealthy. On the far technical horizon, there are alternative procedures that might provide relatively inexpensive mitochondrial transplantation with perfect haplogroup matching, but none of these have so far reached a state of disclosure and development that is appropriate to discuss here.

The choice between the above approaches would fundamentally determine whether age reversal becomes a universal human right or remains stratified by genetic lottery and economic status. In the worst case, regulatory agencies might have to make unprecedented decisions about setting the acceptable mtDNA compatibility standards, with stakes that involve long-term health and human longevity itself.

10.4 The Consequences of Mitochondrial Mismatch

Here, the key unresolved question is: *What are the consequences of having in most body cells a major fraction of mitochondria with a qualitatively different mtDNA sequence than the one with which the recipient was born?* In a blood transfusion, a blood-type mismatch potentially could be damaging or fatal to the recipient. In a mitochondrial transplant, a haplotype mismatch would clearly never be fatal, but it might have unwanted (and presently unknown) downstream consequences. The mismatch issue is presently unresolved, but there are some indications of problems following a mismatched transplant.

Over a time span of around 1.5 billion years, the mitochondrial genome and the part of the nuclear genome associated with mitochondria have co-evolved. In mitochondria, this co-evolution has led to optimized interactions between the proteins encoded by the nuclear DNA (which are synthesized in the cell cytoplasm and imported into mitochondria) and the 13 proteins encoded by mtDNA (which are hydrophobic, would not travel well in water-based cytoplasm, and instead are produced locally to provide parts and support for the mitochondrial ATP production bio-machinery).

The different mtDNA haplogroups represent about 42 major branches and sub-branches in the human mitochondrial haplo-tree, reflecting varying ancestral migrations and adaptations to different environments (e.g., climate, diet, …). These adaptations can show up as subtle differences in mitochondrial function. For example, some haplogroups seem to have adaptations in either maximizing ATP production or maximizing body heat generation. These adaptations are likely to be related to living and evolving in very hot or very cold climates.

When mitochondria from a different haplogroup are introduced into a cell with nuclear DNA that co-evolved along with the mtDNA of a different haplogroup, this finely tuned mito-nuclear compatibility can be disrupted.

This disruption can lead to alterations in ATP production, which is a complex interplay between the nuclear DNA-encoded and mtDNA-encoded proteins that need to fit together. The result of such a mismatch can be reduced ATP production, increased ROS generation, fewer Complex V ATP-producing sites per mitochondrion, or other control and metabolic problems. However, we note that in the cell nucleus, half of the mitochondria-related genes are inherited from the father, so any such maternal-line mito-nuclear gene optimization will necessarily be limited.

Experimental studies with "cybrid" cells (cells with nuclear DNA from one individual and mtDNA from another) and with "mitochondrial exchange" mice have shown that different mitochondrial-nuclear-DNA combinations can affect nuclear gene expression, cellular bioenergetics, oxidative stress levels, and even susceptibility to certain diseases (e.g., atherosclerosis, insulin resistance, heart failure, ...).

Research has also shown that certain mixed combinations of mtDNA variants, particularly those with incompatible mitochondrial haplogroups, can lead to mitochondrial dysfunction, increased ROS, and dysregulation of connective tissue gene expression, potentially contributing to conditions like brain malformation, joint hyper-mobility, and disruption of the autonomic nervous system.

In other words, there are known negative downstream consequences from mtDNA transplantation mismatch, particularly for large-volume transplants. This concern raises an interesting question: Are there "type-O" mitochondrial haplogroups for which the mismatch effects are kept to a minimum? This question remains to be answered.

The precision required for mitochondrial matching would determine both treatment efficacy and accessibility. Detailed matching might prove unnecessary if broadly compatible mitochondrial variants can be identified, but the consequences of a mismatch could range from reduced treatment effectiveness to serious autoimmune reactions. Further research and early clinical experience should be crucial in determining minimum compatibility requirements.

Such research seems to likely reveal that certain mitochondrial functions require precise matching, while others can tolerate significant haplotype variation. This could lead to tiered mitochondrial treatment. At the lowest level would be relatively inexpensive or free transplantation using generic mitochondria with a random haplogroup (essentially what is being done now at low volumes with young plasma replacement and stem cell therapy) or perhaps with a standard popular haplogroup. At the more expensive second level would be mitochondrial transplantation with the same general haplogroup (e.g., Type H or Type H2 or Type H2a or Type H2a1 or …) matched at the high levels, but with the more extended details of the haplogroup ignored, likely resulting in a small and perhaps inconsequential mismatch. At the expensive top level would be mitochondrial transplantation with the perfectly matched mitochondria that are either derived from the subject themselves or from a maternal line relative with the same mtDNA sequence. That option would likely require at least the temporary use of a dedicated bioreactor (see above).

The regulatory framework would need to balance safety and long-term health concerns against treatment accessibility, potentially creating different approval pathways with different costs and levels of matching precision. This would be tricky, since it may require many years to determine just what the long term consequences of a haplogroup mismatch in mitochondrial transplantation really are.

10.5 Impact on Amateur and Professional Sports: Mitochondrial Doping and Performance Enhancement

Competitive sports are already grappling with complex "doping" problems. If generic mitochondrial transplantation soon becomes a standard treatment, widely used in hospital emergency rooms, its spill-over effects to competitive sports will be far-reaching and profound [Wh25]. It will impact everything from levels of athletic performance to anti-doping regulations and detection methods. It will perturb the essential definition of just what "natural" athletic ability is.

Here's my estimate of what is to come. Dr. James McCully is a pioneering mitochondrial researcher at Boston Children's Hospital who participated in the first successful human mitochondrial transplantation to the heart of a dying baby girl and other children. He recently reported (Wh25) that, due to news articles about his ground-breaking work with mitochondria, he has already been approached by athletes seeking his services for mitochondrial infusions to boost their performance levels. Today, mitochondrial transplantation therapy remains very preliminary and experimental, but Dr. McCully says "It seems to have worked in every organ we've looked at."

Amateur sports, including college-level and Olympic-class competitions, already have in place very strict anti-doping regulations and have created an agency to enforce them. On the other hand, professional sports make their own rules, which presently are rather more lenient than the rules that apply to amateur competitions. For example, the National Football League (NFL) presently imposes a 4-game suspension on any offending player for the first offense of a doping violation involving stimulants or steroids. In contrast, an amateur athlete can be banned from their sport for two years for exactly the same violation. The professional leagues often focus on "drugging" substances relevant to their sport, and they often allow therapeutic use exemptions for medical substances, particularly those that are widely used in treating sports-related injuries.

10.5.1 A New Class of Sports Doping: The Rise of "Mitochondrial Doping"

The World Anti-Doping Agency (WADA), which enforces anti-doping rules for amateur sports, has already put in place a framework that might classify mitochondrial transplantation as a prohibited method. The WADA Prohibited List includes "Gene and cell doping" (Category M3), which specifically bans the non-therapeutic use of cells, genes, genetic elements, or the modulation of gene expression to enhance performance. A court of law would probably have to decide if increasing the *number* of one's own mitochondria already present in the body, which are not precisely a cell, a gene, a genetic element, or a gene expression modifier, would be included and banned under this prohibition. Therefore, mitochondrial transplantation might or might not fall under this category and be prohibited.

Should it be prohibited? By infusing a concentrated dose of healthy mitochondria into the athlete's bloodstream or injecting it into specific sites (e.g., arm or leg muscles), it is expected that an athlete could boost his or her strength and aerobic capacity, could delay the onset of fatigue, and could speed up the recovery time after tiring exertion. All of these are highly sought-after advantages in competitive endurance-based and power-based sports. Thus, considering its likely effects, the mitochondrial transplant procedure might be considered be a form of "cellular doping" that manipulates and enhances the body's fundamental energy systems to obtain a huge competitive edge.

10.5.2 The Detection Dilemma: An Arms Race at the Cellular Level

Assuming that mitochondrial transplantation was indeed prohibited, one of the most significant challenges for anti-doping authorities would be to establish a valid and unambiguous determination of the fact that mitochondrial transplantation had in fact been done. This would be particularly difficult if the athlete's own (autologous) mitochondria were used in the transplantation process.

10.5.2.1 Autologous Transplantation

The process starts by obtaining the stem cells to be amplified. If the stem cells used in the production of mitochondria were taken from an athlete's own body (e.g., muscle tissue, blood cells, bone marrow, …), amplified in a bioreactor, and then the extracted mitochondria were coated to suppress immune response and infused into the athlete's bloodstream, there would likely be *no foreign genetic material* that would be detectable. The average mitochondrial DNA copy number might be significantly increased. This is measurable in principle using droplet digital PCR (see Chap. 5, Sect. 5.10 above), but establishing a baseline for that indicator and using it to conclusively prove that there had been illicit enhancement would be quite challenging. We note also that in current cases of athletes engaging in "blood doping", the presence of the plasticizer DEHP (*di-2-ethylhexyl phthalate*) in blood and urine samples has been taken as a strong indication of a possible infraction. If stem cells and/or mitochondria were stored in plastic medical containers containing that additive before infusion, the same test might serve as an indication of mitochondrial doping.

Otherwise, there appear to be no testable indications that a mitochondrial transplant had ever taken place. Standard sports doping tests, which often look for synthetic substances or their metabolites, would be completely ineffective. Detecting such a subtle biological enhancement might be effectively impossible, or at best would require a new generation of sophisticated and potentially invasive testing methods, possibly involving repeating time-dependent short-cycle monitoring of an athlete's biological status to look for subtle changes with time in cellular markers or metabolic byproducts.

10.5.2.2 Allogeneic Transplantation

Alternatively, the transplanted mitochondria might have come from the stem cells of a random or selected donor, perhaps because these were more readily available for immediate use, for example, from a hospital emergency department's supply of "banked" generic mitochondria. This transplantation would, in principle, produce a detectable condition. Mitochondria have their own DNA, a small ring of 16,569 base pairs that can be readily sequenced and its structure recorded. The sequencing of an athlete's mitochondrial DNA after a donor-based transplant would be likely to show the presence of heteroplasmy, i.e., two distinctly different mitochondrial DNA haplogroups (inherited mutation patterns). To avoid such detection, determined athletes and their supporting team physicians might choose mitochondria with a matching top-level haplotype or seek out a donor with a matching mitochondrial DNA haplotype, e.g., from a maternal-line relative, to minimize the chances of such detection.

We note that any detection of such a haplogroup mismatch is not likely to be accomplished rapidly. Such mitochondrial DNA genetic analysis, currently available from genetic genealogy firms for a few hundred dollars, involves PCR amplification followed by full nanopore sequencing or bio-chip analysis, and this process normally requires several weeks. Further, athletes whose parents had used (or can claim to have used) *in vitro* fertilization involving donor mitochondria to avoid genetic disease might already have two haplotypes of mitochondrial DNA in their systems and would certainly test positive, but such a result would have nothing to do with any performance-enhancing doping. Similarly, if an athlete had received a bone-marrow transplant, heteroplasmy might be detected. Thus, this mitochondrial doping scenario sets the stage for a new technological "arms race" between those seeking an illicit advantage in sports and the anti-doping establishment. This competition should push the boundaries of scientific detection and forensic analysis.

10.5.3 The Blurred Line Between "Therapy" and "Enhancement"

Thus, the expected widespread availability of mitochondrial transplantation in emergency medicine in a few years should create a significant ethical gray area in sports. An athlete who suffers a legitimate injury, e.g., a severe muscle tear or a brain concussion, might need to receive mitochondrial transplantation as a standard part of their medical treatment. This therapeutic intervention could, as a side effect, greatly enhance their baseline athletic capabilities upon their return to competition.

This scenario raises critical questions for anti-doping authorities:

- *How can a legitimate therapeutic use of mitochondrial transplantation be distinguished from a deliberate attempt to enhance performance?*
- *Would athletes who receive mitochondrial transplantation for a valid medical reason have an unfair advantage over their competitors?*
- *Should there be a "stand-down" period after therapeutic mitochondrial transplantation before an athlete could return to competition? If so, for how long?*

The existing practice in sports has been to provide a Therapeutic Use Exemption that allows athletes to use certain prohibited substances for necessary medical reasons. This will become an incredibly complex territory to navigate.

10.5.4 Aging Sports Superstars

A common theme in major professional sports today is the aging "superstar" player who could greatly benefit professionally and financially from squeezing in just one more successful playing season. The advent of mitochondrial transplantation could profoundly reshape the landscape of the aging star players of major professional sports. For the senior superstars of the NFL, MLB, NHL, NBA, and international soccer, mitochondrial transplantation represents more than just a theoretical advantage. It offers the tantalizing prospect of extending their prime performance, rewriting the record books, and dramatically altering the economic and ethical fabric of their leagues.

So ultimately, the arrival of mitochondrial transplantation in the prosperous world of professional sports is very likely to be a matter of "when," not "if." It represents the next frontier in the ongoing competition between natural human ability and technological intervention. For injured players, it offers fast recovery. For aging superstars, it promises a second wind, a chance to defy Father Time and Mother Nature. One can imagine the aging star quarterback employing his own longevity-medicine technical staff, who would set up a private bioreactor lab that operated quietly and continuously to amplify his own stem cells, to supply his own mitochondria for his secret monthly mitochondrial transplant infusions, all at a net cost that was a rather small fraction of his large annual salary.

For the professional leagues that pay and play these star athletes, this promises to trigger a cascade of regulatory, ethical, and existential challenges that would force them to decide which option they value most: (1) the sanctity of their athletic ethics and records or (2) the spectacle of their greatest stars shining brighter and longer. The professional leagues all make their own rules, which have been known to bend before the wind of economic pressure. In the long run, spectacle and ticket sales and rising profits seem likely to prevail over ethics and records.

The potential for mitochondrial doping strikes at the heart of the "spirit of sport," a core principle of the World Anti-Doping Code that emphasizes fair play, honesty, and the pursuit of human excellence in amateur sports without the aid of artificial enhancements.

The inevitable introduction of a technology that can fundamentally alter an athlete's cellular physiology raises profound questions about the future of competitive sports. If success becomes increasingly dependent on access to cutting-edge and very expensive medical technologies, it could create a two-tiered system of "haves" and "have-nots," in which financial resources rather than natural talent and dedication become a primary determinant of victory.

Thus, while mitochondrial transplantation holds immense promise for the field of medicine and particularly of longevity extension, its potential for becoming a standard emergency room procedure will cast a long shadow over the world of competitive sports. It will force a comprehensive re-evaluation of anti-doping strategies, ignite complex ethical debates, and potentially redefine what it means to be a twenty-first-century athlete. The looming challenge for sports governing bodies will be to proactively address this emerging technology before it becomes the next frontier in the ongoing battle against doping in sports.

(*Note:* A version of Sect. 10.5 above appeared as a "The Alternate View" column by the author in the March–April-2026 issue of ***Analog Science Fiction and Fact Magazine.)***

10.6 The Military Mitochondria Spinoff

If the use of mitochondrial transplantation becomes important in emergency medicine and in enhancing the performance of athletes, its military applications will not be far behind. Stocks of generic mitochondria for emergency treatment of battlefield injuries will become common. Soldiers might be specifically chosen by haplotype to match that of the mitochondrial supplies that were available on a particular battlefield.

In previous US engagements, the military has acted in quite the opposite mitochondrial direction. In the 1990–91 Gulf War, concern about the possibility that anthrax might be used as a biological weapon by Saddam Hussein in Iraq caused the US Army to pre-inoculate its soldiers with the anthrax-specific antibiotic drug Cipro before they went to overseas. As it turns out, Cipro acts by interfering with the replication of the DNA ring in the anthrax bacterium, and it also does the same thing to mtDNA, killing many of the soldiers' cellular mitochondria and possibly damaging the mtDNA of their surviving mitochondria. The result is Gulf War Syndrome, which has been shown to be largely the result of mitochondrial damage incurred during the Gulf War deployment. (We note that cipro-family antibiotics are still widely used to treat urinary-tract inflections in women.)

In any case, mitochondrial infusion offers the potential for soldiers to heal more effectively, improve recovery from injuries, and avoid chronic war-related disabilities in veterans. This could be achieved by regular mitochondrial infusions coupled with regular measurements to monitor the results. A large military investment in dedicated bioreactors can be expected, once the strategic value of mitochondrial therapy is realized.

10.7 Noted Science Fiction Authors and Aging Remedies

In the context of extending human longevity, it is interesting to consider the extrapolations that noted science fiction authors have already used in their works of fiction. In ***Methuselah's Children*** (1941) Robert A. Heinlein first described the Howard family, a group who had extended their lifespans by selective breeding and ultimately needed to escape Earth because of governmental demands that they must reveal their "longevity drug secrets". In their relativistically time-dilated absence, government-funded research yielded lifespan-extending procedures, and these subsequently enabled Heinlein's Howard-descended protagonist Lazarus Long, as described in his novel ***Time Enough for Love*** (1973) and elsewhere, to live for more than 2,000 years. We note that Long was *forced* to undergo longevity enhancement treatments against his will.

Larry Niven, in his ***Ringworld*** (1970) and the other books of the Known Space series, described *boosterspice*, a not-well-specified lifespan-extending drug used by several protagonists. There were hints that it may have originally been derived from a virus present in the Tree-of-Life plant associated with the immortal post-human Pak Protectors.

Bruce Sterling, in his 1996 novel ***Holy Fire,*** while the novel does not mention the word "mitochondria", describes a plausible near-future civilization in which human aging has been eliminated by means of radical life extension technologies that are highly intrusive and experimental, placing the recipient in a rejuvenation tank for many months during invasive treatment. His protagonist, Mia Ziemann, a moderately wealthy 94 year-old woman, decides to undergo these transformative medical procedures and have her youth restored, effectively reversing the effects of many decades of aging. She was provided treatment because she was among the few who had lived virtuous, risk-averse lives and had accumulated enough financial resources to afford the expensive procedure. The rejuvenation technology involved advanced biomedical interventions, employing telomere, senolytic, genetic, cellular, neural, and organ regeneration therapies. Her rejuvenation was not just physical. It awakened dormant hormones and added tissue to her brain, producing psychological and existential transformations that propelled Mia into a compelling post-human psychological condition and stimulated her subsequent rule-breaking adventures.

Damon Knight's 1957 short story "The Dying Man" describes a civilization in which human aging was arrested and nullified through a vaguely described pre-birth genetic treatment, eliminating the underlying biological processes that produced human aging. This procedure became universal in the story's future society, rendering humans effectively immortal unless killed by accident or violence. The protagonist, Dio, is the last person for whom the aging treatment has failed, making him unique as the only mortal in a world of immortals. His progressive age-degenerating condition is widely publicized and viewed by contemporaries with a mix of curiosity, pity, and reverence. He becomes the icon for the bygone era of the human race, when age degeneration and death were inevitable and inescapable.

Many other science fiction writers have used characters with greatly extended lifespans, often achieved by random mutations or using miracle drugs and pills, systematic organ transplantation, long dormant stays in "rejuvenation tanks", or mind transfer to lab-grown body-duplicate clones. These extrapolations often provide the basis for investigating in detail the societal, economic, and ethical consequences of extended human longevity.

Of course, from our present point of view it is quite understandable that most of these plot lines were well "off the mark" in describing how human age reversal might actually be achieved. Science fiction should (but does not always) follow from the growing understanding produced by science. Our present scientific understanding of what human aging actually is and our insights into how it might be arrested or reversed are very recent. We look forward to new works of science fiction that actually get the science right while telling interesting longevity-related stories.

10.8 The Advantages of Extending Human Longevity

In this section, we want to examine in some detail the many benefits that follow from extending the human lifespan through large volume mitochondrial transplantation and other strong anti-aging interventions.

10.8.1 We Have Already Been Extending the Human Lifespan

In 1900, the human life expectancy at birth in the USA was 48.2 years, and in 1925 it was 58.2 years. That was a 10 year gain in life expectancy in just 25 years. In 1934 (the year I was born) it had inched up to about 60 years. Today it is at 79.6 years and would be even higher if not for the late twentieth century's Opioid Crisis and the megadeaths from the recent COVID pandemic. In other words, since 1900, the human race has increased its average lifespan by more than 30 years or about 65%.

Did this cause massive societal disruptions? No! We adjusted quite nicely, and there was never any serious discussion that I can recall about the negative impacts of the average person living longer. Large volume mitochondrial transplantation is a medical treatment, like hundreds of other medical treatments that we have already brought into play in the last hundred years. We have adjusted to major lifespan changes in the past with little disruption or concern.

Further, we are already extending human life, whether it is with chemotherapy to treat cancer, insulin for protection from the fatal effects of diabetes, heart transplants to prevent untimely death from cardiovascular disease, or organ transplants to deal with otherwise fatal organ failure. They are all medical treatments currently in use that extend the longevity of the recipients.

10.8.2 Extended Longevity Frees Humanity to Accomplish More

What new ideas and theories would Einstein have developed if he had another 20 years of life? What would Picasso have painted if given 20 more years? What music would Mozart or Beethoven have composed? What novels would Hemingway have written?

We all may soon have the opportunity to learn new skills, to pursue new ideas, to be more productive, to do more with our lives. We are not all Einsteins and Hemingways, but on the whole, in our productive years, we have made significant contributions to society. If longevity extension is available, we should take advantage of it!

10.8.3 Extending Longevity Is Long Overdue

> Fifty years…twenty of them he is a child, he is well past thirty before he is skilled in his profession. He is forty before he is established and respected. For not more than the last ten years of his fifty has really amounted to something…And now, when he has reached his goal, what is his prize? His eyes are failing him, his bright young strength is gone, his heart and wind are 'not what they used to be'. He is not senile yet, but he feels the chill of the first frost. He knows what is in store for him. He knows—he knows!
>
> Robert A. Heinlein, ***Methuselah's Children***, 1941

Today is the right time to extend longevity. In fact, it's long overdue. Society has become extremely complicated. Many professional careers don't even get started until the person is 30 years old. When they are 65, most are supposed to retire, and society loses the benefits of their decades of training and experience.

And, as discussed in Appendix B below, the retiring 65 year-olds will face the prospect of rapidly declining fitness and health, as the deletion mutations in their mtDNA begin to increase exponentially, doubling about every 3 years, driving their aging bodies to face a growing shortage of ATP energy, a shortage that will be shutting down the energy-expensive cellular repair mechanisms that keep them young. It is not rational to tolerate this situation, when fairly simple procedures are becoming available that can correct it.

The enabling technologies of smartphones, GPS, the Internet, and AI are upon us, and the even more enabling technologies of quantum computing, self-aware general AI, neural implants, and direct mind control of devices loom not far in the future. We are beginning to have physical and intellectual powers that dwarf those of past generations, yet we are stuck with the same weak, easily-damaged, short-lived bodies. That situation is ripe for a change.

10.8.4 It's Not Just for the Elderly

Millions of young people suffer from damaged mitochondria, prematurely aging their systems. Every year, thousands of children are born with prematurely aged mitochondria and die after long, painful, and extremely expensive attempts to keep them alive. Military veterans and others can be damaged by chemical exposures that leave them prematurely aged. The veterans with Gulf War Syndrome show definite signs of premature mitochondrial damage. Astronauts encountering radiation in space from cosmic rays have been prematurely aged, and there are significant concerns about potential radiation exposure from cosmic rays and solar flares during any Mars mission of many months. Any mitochondria-based treatment that is developed for emergency room use and the elderly will apply equally to all of these other groups.

In 2020, adults of ages 85 and older, who are 2.5% of the population, had a *per capita* yearly health care cost of almost $36,000, over 8.5 times higher than the health care cost for an equivalent child. We can anticipate huge cost savings in medical expenses from the introduction of mitochondrial treatments. Care for the elderly currently comprises more than 50% of total healthcare spending worldwide. If we can reduce the incidence of age-related diseases or even delay them for a decade or so, that will have an extraordinary impact on reducing overall worldwide healthcare expenditures.

10.8.5 The Cost Is Comparable to Other Medical Treatments

Cell phones were once only for the rich. The Internet was originally designed to be used only by large government and defense-related organizations and elite science laboratories, and now we all have it. Of course, the cost of mitochondrial transplantation will be high initially, just as is the cost of most new high-tech products. We will need to gain the knowledge and investment to mass-produce it for global distribution.

Treatments like that used for spinal muscular atrophy cost over $ two million per dose. A heart transplant costs about $ one million. CAR-T cell therapy, a cutting-edge anti-cancer treatment, costs around $500 k per patient. Big Pharma currently spends hundreds of millions to develop and test each new drug, and then passes the costs on to consumers, mainly those in the USA. As time goes on and production and surgical techniques improve, all new drugs and procedures become less expensive (unless patent holders keep the cost artificially high.) Already, there are innovative ideas for greatly reducing the cost of mitochondrial production.

One stated goal from the beginning of the mitochondrial transplantation project has been the vision of a future in which huge factories are pumping out mitochondria by the ton, and their product is made available to everyone over the age of 55, or perhaps even younger. Think of renewing your mitochondria as similar to taking the newly available "Vitamin M" or getting your annual flu shot.

The technique is brand new and still largely untested (although all tests so far with animals and humans in restricted circumstances have been very promising). Nothing has yet been proved. Many uncertainties and unknowns remain. Large amounts of research still need to be done. But we must take the first steps.

10.8.6 Longevity Insurance Versus Life Insurance

Assuming that the costs of mitochondrial transplantation will remain high for some time, the long-term life horizon will encourage individuals to save more as they work for more decades, viewing their long-term goals and seeking to accumulate funds to implement a comfortable extension of their longevity. This increased overall savings should fuel greater investments in the economy, leading to capital investments that deepen and further economic growth.

One can imagine the rise of a new financial industry offering "longevity insurance" analogous to the present life or health insurance. Participants might make monthly premium payments to such an insurance policy for perhaps 20–40 years during full employment, to receive at policy maturity (provided one is alive to receive it) the full-blown comprehensive autologous mitochondrial transplantation treatment using a dedicated bioreactor or the equivalent for full age reversal. This would be implemented when one reached some specific "rejuvenation age" (rather than a retirement age). The cost of the monthly payments might depend on the extent of the future treatment, the degree of haplogroup matching, and the time until rejuvenation age, with less restrictive matching and a longer time span between rejuvenations allowing for lower payments. In this way, even if the eventual cost of rejuvenation by mitochondrial transplantation turns out to be very high and remains there, the treatment could still be readily accessible to the prudent middle class.

One of the most significant immediate benefits of such insurance might be the reduction of pressure on public and private pension and retirement systems. If people worked for another 10 years or more, they would contribute to these systems for a longer period and draw later and perhaps for a shorter period (proportionately), improving the financial sustainability of retirement programs.

The extended lifespan would also create and expand markets for products and services aimed at older recipients, a wealthier, and more active demographic. This includes healthcare, specialized housing, leisure activities, education, and financial products tailored for longer retirements. This "silver economy" would become a more powerful driver of economic activity.

A spin-off industry based on extended longevity technology might be its application to pets and to sports animals, particularly dogs, cats, and horses, all of which presently have dismayingly short lifespans. Wealthy celebrities have already invested large sums to have their beloved pet dogs cloned, in an effort to give the pet a new life of a sort. Racehorses are a large financial investment, with a large payoff for the winners that could be extended by giving champions a longer time to pursue active racing competition. Further, the mitochondrial doping of racehorses and sports dogs is likely to develop along with the mitochondrial sports doping of human athletes described above in Sect. 10.5.

In summary, assuming that a lifespan increase meant that people remained healthier for longer, it might significantly reduce the burden of age-related chronic diseases and the large associated health care and elder care costs of later life. This shift from "living longer, being sicker" to "living longer, being healthier" would spark a major economic boon.

10.8.7 Economic Impacts: Longer on the Job, Better Use of Experience

Instead of assuming that a healthy and productive working life will be extended for a long but indefinite period, let's here consider a simpler case: Suppose the accumulated biomedical breakthroughs allowed the human lifespan and healthspan to increase by exactly 10 years, so that the work duration until retirement would increase by a similar increment (e.g., to age 75 instead of the present 65 or less). What would be the social and economic impacts of this development, in both benefits and problems?

Economist Scott [Sc24] has described in detail the opportunities that extension of human longevity creates. With people working for an additional 10 years, the global labor force would expand significantly (~ 20%).

This larger workforce, especially if accompanied by continued good health and cognitive abilities (as suggested by some research indicating that "70s are the new 50s"), would lead to significantly increased overall productivity and economic output. Older workers often provide valuable experience, deep institutional knowledge, and a strong work ethic, which can boost efficiency and stimulate innovation.

A longer working life would directly contribute to the national GDP. The International Monetary Fund (IMF) has already projected that the present trend toward healthier aging, leading to increased labor supply and human capital, should add approximately 0.4 percentage points per year to global GDP growth over the next few decades. The kind of age reversal we are considering here would do far more than that.

A longer lifespan and work duration should motivate greater investment in education and skills development throughout a person's life. Individuals would have more time to acquire new knowledge, to track production improvements, and to adapt to technological changes, leading to a more skilled and adaptable workforce.

Some might be concerned that the older workers, holding on to established jobs, would "crowd out" younger generations from the labor market. However, given the currently declining worldwide birth rate and the population reduction produced, it is more likely that the accompanying economic growth from working longer would compensate and create new jobs for all ages. Nevertheless, employers might face challenges in adapting workplaces to accommodate an older workforce, including issues of physical demands, technological proficiency, and potential biases against older workers. Again, it's important to remember that for the past century our economy has been adapting quite successfully as average lifespans have extended by about 30 years, and the nature of labor has dramatically changed, from agricultural work to factory work to information work. It seems most likely that in the future our economy will continue to seamlessly adapt to longer lifespans and related changes.

While older workers bring experience, there is a need to ensure their skills remain relevant in a rapidly evolving job market. Vigorous implementation of continuous re-skilling and lifelong learning programs would be crucial, and their funding could be a challenge. While the longevity economy presents opportunities, there could also be shifts in overall consumer demand away from goods and services typically consumed by younger demographics, requiring businesses to adapt.

While a significant increase in lifespan and work duration presents tremendous potential for economic growth, increased productivity, and the sustainability of social welfare systems, realizing these benefits will require policy decisions that promote healthy aging, lifelong learning, flexible work arrangements, and equitable access to opportunities across all age groups. Intelligent legislation by local and national government bodies that foresees possible problems and proactively deals with them would likely be needed. Careful planning and adaptation will ensure that the challenges of intergenerational equity, healthcare costs, and workplace dynamics support positive economic impact.

10.8.8 Societal Impacts: Population, Jobs, Annuities, Tenure, and Seniority

It has been widely reported that in the next decades of the twenty-first century, the world civilization is on the brink of a "depopulation explosion", because the worldwide birthrate, particularly in most developed countries, is falling well below the 1:1 replacement level required to balance deaths with births. For example, the number of children for the average woman in the USA is 1.62, in Canada it is 1.26, in Mexico it is 2.0, in the UK it is 1.41. In France it is 1.62, in Ireland it is 1.5, in Germany it is 1.35, in Italy it is 1.20, and in the European Union as a whole it is 1.38. In China it is 1.05, in India it is 2.0, and in Pakistan it is 3.3. It needs to be 2.1 to avoid a population decrease. Immigration is presently holding most national populations constant or growing, but this is not a popular or permanent solution. Unless this trend is corrected or compensated for, the world and national human populations are destined to decline steeply, leading to many undesirable consequences.

However, if we are also on the brink of a massive increase in human longevity, the age-related death rate should also steeply decline, which would tend to compensate for the decreasing birth rate. It is not clear how these opposing trends will play out. It is possible that they could be guided to a balanced and relatively constant human population of the planet, leading to stability and easing the ecological burdens. However, even if that were accomplished, the average age of the population would continue to increase for some time, hopefully accompanied by a healthspan that similarly increases.

Faced with a workforce that would be much slower to age out and retire, many traditional employment structures would need to adapt. Assumptions about career progression, standard retirement, and generational job turnover will change. Tenure systems in academia would face existential crises, when professors could potentially hold tenured positions for centuries rather than decades.

In government, the existing seniority systems of the United States Senate and House of Representatives already result in relatively "permanent" chairs of key committees, to the detriment of a healthy and progressive government. Similarly, Supreme Court Justices seldom retire. These existing problems might become even more severe with extended longevity. Imposing strict term limits on all political and judicial offices might offer a feasible remedy, if such limits could be implemented without obstruction from the powerful individuals affected. Some requirement that departing term-limited politicians could not move to positions related to government, e.g., lobbying or government-related consulting, might also be essential.

Retirement as a social institution might be changed, forcing restructuring of private pension systems, the federal Social Security program, and related age-based benefits. Workers might cycle through multiple careers over extended lifespans, transitioning to a new job or profession every two or three decades. This would likely require continuous re-education and adaptation systems accommodating students of all ages.

If there was true rejuvenation to a much younger profile, the concept of "youth" itself might need to be redefined when individuals could maintain physical and cognitive youth for centuries. Traditional life stages of education, career building, family formation, and retirement would need fundamental re-conceptualization for populations that might have lifespans of centuries.

10.8.9 Preparing Improved Humans for Space Colonization

Space colonies on the planets, moons, and asteroids of the Solar System and in constructed space habitats in Earth orbit and elsewhere, if and when they become feasible, might make the transition from a distant aspiration to a real possibility as human beings can become more durable, better adapted to low gravity, longer lived, and more resistant to radiation. The technology enabling age reversal could thus enable humanity's expansion beyond Earth. Application of this technology could be an effective response to the fundamental problem that troubles this optimistic man-in-space scenario: *humans are presently too fragile to have anything resembling a healthy life in space.*

Space, for us humans, is a very hostile environment. There's no air for you to breathe, and "no one can hear you scream." There's also little or no gravity, the lack of which over a few months is known to cause your muscles to degenerate and your bones to lose mass. Further, outside the Earth's protective geomagnetic field and shielding atmosphere, your body will be irradiated by much more ionizing radiation, some of it heavily ionizing. This will damage or kill the cells of your body, will prematurely age your mitochondria, will produce dangerous nuclear and mitochondrial DNA mutations that your future children may inherit, and will increase your chances of developing cancer after a few decades.

The SF writer Charles Stross, in his novels in the ***Saturn's Children*** universe and particularly in his short story "Bit Rot", envisioned a race of post-human android servants that had been engineered by humanity to better withstand the hostile environment of space, including working in vacuum without a spacesuit, better resisting radiation damage, enduring long periods of zero gravity without degeneration, and living longer for extended interstellar trips. In the Stross scenario, some time in the twenty-third century unmodified humanity will go extinct due to our fragility and impulsiveness, and our former humanoid servants will take over the jobs for which they are better equipped: exploring space and populating the habitable star systems of our galaxy.

However, while re-engineered humanoids might be an elegant solution to the fragile man-in-space problem, realistically, we must work with *the humans that are presently available*. This raises the question of whether there are any technological workarounds to deal with the space-related frailties of humanity.

10.8.9.1 Radiation Damage

For NASA, the problem of space radiation damage has no known techno-fix. Our Sun, particularly at times of solar flares, spews out floods of fast electrons and protons that make the colorful Northern Lights on Earth and represent a serious radiation hazard for space travelers. Moreover, galactic and extra-galactic cosmic rays include a fraction of very energetic, highly-charged atomic nuclei, frequently the iron nuclei from supernovas, that heavily ionize in matter and create a penetrating radiation shower in the best radiation shielding. These energetic bare nuclei ionize so strongly that, if they encounter the mitochondrial or nuclear DNA of a cell, they will almost always break both bonds of the double helix. In nuclear DNA, such a break makes the natural repair process more problematic. In mtDNA, a double break is not repaired at all, and it frequently leads to exponentially increasing deletion mutations (see Appendix B below).

The standard international unit of radiation dosage is the *sievert* (or Sv), defined as one joule of ionizing radiation energy deposited in one kilogram of body mass. A sievert represents a seriously large radiation exposure. A dose of about 4.5 Sv will kill about half of a group of exposed humans in 30 days. More typical exposures are measured in millisieverts (or mSv), one thousandth of a sievert. For example, the average human living on the Earth's surface at sea level will receive a yearly dose of about 3.6 mSv, a CT scan delivers a dose of about 8.5 mSv, and a Department of Energy radiation worker is allowed to accumulate a yearly dose of 20 mSv.

A Mars colonist would receive a yearly dose of about 234 mSv if living unshielded on the surface of that planet. A 234 mSv dose is not lethal, but it damages mtDNA, greatly increases the likelihood of mutations in the children produced by colonists, and it increases the chances of cancer in later life. To put it another way, unshielded life on Mars will deliver a dose of ionizing radiation that is 65 times larger than that of the average Earth resident and 12 times larger than that allowed for a DOE radiation worker. That does not include the larger but variable radiation exposure on the initial trip to Mars. It is certainly enough to warrant a great deal of concern about the health and long term survival of Mars colonists.

The cells of a living organism damaged by ionizing radiation can take three paths: cell repair, cell senescence, and cell death. For very large radiation doses, the dominant effect is cell death that brings with it the symptoms of acute radiation poisoning: immediate hair loss and low blood pressure, nausea and bloody vomiting in 10 minutes after exposure, bloody diarrhea and fever in one hour, severe headache in two hours, soon followed by death. For milder exposures, the outcome depends on the dose and kind of radiation. Electrons, muons, x-rays, and gamma rays tend to produce single breaks in DNA strands that, if they don't pile up, can be fixed by the ever-present internal cell repair mechanisms. Protons, alpha particles, and heavier nuclei have high ionization densities that produce more damaging double DNA breaks. That radiation damage might be in a "junk" DNA region where it would have little effect, but some of the double-break radiation damage, particularly to the mtDNA, will inevitably render the cell non-functional.

In humans, even without radiation exposure, the cells in the high-damage locations of the body are replaced frequently. For example, neutrophil white blood cells, the front-line warriors in the immune system, are replaced every 7 hours, blood platelets are replaced every 8 days, intestinal cells are replaced every 10 days, skin cells are replaced every month, red blood cells are replaced every 4 months, and liver cells are replaced every year. In case of mild radiation exposure, cell death is perhaps preferable to cell senescence, as long as it does not add too much to the normal rate of cell replacement.

Ionizing space radiation is normally not intense enough to produce massive cell death in exposed humans in space. Rather, cell damage accumulates over a period of time, with the rate of damage accumulation much larger than in any environment on Earth. That accumulated damage mainly takes the form of senescent and pre-cancerous cells. With the added body burden of senescent cells, the space-traveling humans will have reduced fitness, premature aging, and a much greater cancer risk.

At present, the only remedy for space radiation exposure that has been seriously considered by NASA is the use of in-flight shielding, which requires significant mass that must be transported to orbit and beyond. It's envisioned, for example, that if a Mars mission carries with it a large quantity of water for consumption and possibly for propulsion reaction mass, the crew might be housed behind water-filled walls to reduce space radiation exposure, and there might also be smaller, highly shielded regions to which the crew could retreat in the event of a major solar flare. Requiring and planning for such shielding greatly complicates any manned-mission design.

Fortunately, the emerging technology of large volume mitochondrial transplantation offers a better alternative for dealing with the problems of space radiation. The radiation-induced loss of cellular function in the bodies of astronauts would be largely due to direct damage to their mtDNA. Such damage could be sporadically or continuously treated in flight by including in the mission a small highly shielded cryo-bank in which were stored (with around a 1 year self life) a few kilograms of the preserved mitochondria of the onboard astronauts. For longer missions with more available space, the onboard equipment might include a compact bioreactor that would continuously amplify stem cells and produce mitochondria for radiation treatment during the mission. Onboard regular testing of mitochondrial integrity, coupled with this treatment mechanism for dealing with radiation-induced mtDNA damage, should effectively neutralize the space radiation damage problem.

10.8.9.2 Microgravity Degeneration

Another problem for fragile humans is that if you spend a few months in microgravity, over that period your body will suffer musculoskeletal degeneration. In particular, you will lose bone density. Your weight-bearing bones (spine, hips, legs) will lose about 1.5% of their mineral density per month in microgravity, leading to increased fracture risk. Your muscles will atrophy, with those used for posture and movement weakening rapidly due to lack of gravitational resistance. You will suffer joint stiffness. Reduced joint use in microgravity can cause joint discomfort and reduced mobility. Without compression from gravity, your spine will stretch, increasing your height by up to 5 cm and often causing back pain.

Further, there are negative effects in the cardiovascular system. Fluids shift toward the head, giving astronauts a "puffy face, skinny legs" look. After returning to gravity, astronauts may faint or feel dizzy, because in microgravity their cardiovascular system has adapted to the absence of a need to pump blood upward against gravity. The heart muscles can shrink and lose efficiency. The neuro-vestibular system is also affected. After return, astronauts have experienced disorientation and nausea at first, due to mismatches between the vestibular system (inner ear) signals versus visual input. Further, astronauts often stagger as though intoxicated until the vestibular system re-adapts. Increased intracranial pressure in spaceflight can flatten the back of the eyeball, swell the optic disc, and induce farsightedness. Some astronauts have experienced lasting vision impairment. It was also found that dormant viruses like herpes and shingles can reactivate, and immune responses may be blunted, making returning astronauts more susceptible to such infections.

There were also other observed effects, including a tendency toward kidney stones from increased calcium in urine, possibly correlated with bone mass loss. There was also a reduction in red blood cell mass and oxygen delivery (space anemia), disrupted circadian rhythms, altered sleep cycles, skin thinning, slower wound healing, and hair loss. MRI studies showed shifts in brain structure and circulation from prolonged microgravity. This is all well documented from the experience of Space Shuttle and International Space Station astronauts.

Interestingly, for the focus of this book, there is evidence from mouse-in-space animal studies that mitochondrial degeneration following an extended stay in microgravity is implicated in many of the symptoms described above [Si20]. Evidence of altered mitochondrial function and DNA damage was also found in the urine and blood metabolic data compiled from the ISS astronaut cohort and NASA Twin Study data, indicating that mitochondrial stress and degradation are consistent results of extended spaceflight.

The observed degeneration of mouse and astronaut mitochondria in microgravity suggests that the conditions of microgravity are reducing ATP production and creating a cellular energy shortage similar to old age, and that the body is responding by shrinking the little-used high-energy-use body parts, resulting in the negative microgravity symptoms described above.

These results raise an interesting question: *If one maintains the quality of astronaut mitochondria by transplantation during extended stays in microgravity, does this reduce or eliminate the observed symptoms of bodily degeneration in microgravity?* The answer to this question is unknown. The International Space Station is scheduled to be retired in five years, by the end of 2030. It would be highly desirable to perform animal or astronaut tests of the microgravity effects of mitochondrial transplantation before that occurs. The resulting information might provide the key to preparing humans for future space travel and colonization.

10.9 Scanning for the "Singularity" on the Future Event Horizon

Many speculative writers and futurists, notably Charles Stross and the late Vernor Vinge, have already warned us of the coming "singularity" predicted to arise from the convergence of longevity extension technology, self-aware general artificial intelligence, quantum computing, genetic engineering, and space exploration and expansion. Severe difficulties might result if these trends simultaneously rise to a crescendo and are managed poorly. However, if managed well they could together create a period of unprecedented opportunities for human civilization, along with some risks. Longevity-enhanced humans with centuries of accumulated knowledge, experience, judicious use of artificial intelligence, and accumulated wealth for intelligent investments could accelerate the expansion of knowledge and technological development exponentially, potentially achieving breakthrough capabilities in physics, chemistry, biology, computing, and space travel within decades rather than centuries. The ultimate question facing humanity is how to ensure that the coming general age reversal represents an evolution toward a higher form of existence, so that the coming age-reversal revolution leads to the unprecedented flourishing of humanity.

10.10 Conclusion

As the legendary Ned Ludd demonstrated in 1779 by initiating the violent eighteenth and nineteenth century Luddite protests against the rise of industrial-scale textile machinery, there is an unfortunate natural human tendency to regard all technological progress as intrinsically dangerous, undesirable, and disruptive. There are also many historical examples of bad industry-management behavior that tend to reinforce such sentiments. Nevertheless, the broader view of recent history shows conclusively that technological progress, when intelligently implemented and managed, is an essential part of human nature and behavior, and that it is not only inevitable but also highly desirable and beneficial. A comprehensive treatment for human aging, looming only a few years away in our immediate future, will be one of the most far-reaching technological advancements in human history, offering the promises of greatly extended human life and health and rapid progress on many fronts. It may require changes in existing societal structures, economic systems, and ethical frameworks, but the result will likely be a very large qualitative improvement in the overall state of human civilization. The transition will present many challenges, but if these are intelligently managed, it should usher in a new Golden Age of unprecedented human potential and achievement.

Clearly, in the not-too-distant future, we are destined to "live in interesting times."

References

[Ad22] A. Adlimoghaddam, T. Benson, and B. C. Albensi, "Mitochondrial Transfusion Improves Mitochondrial Function Through Up-regulation of Mitochondrial Complex II Protein Subunit SDHB in the Hippocampus of Aged Mice," ***Molecular Neurobiology 59*** (10), 6009–6017 (2022); https://doi.org/https://doi.org/10.1007/s12035-022-02937-w.

[Mi25b] "MITOMAP—A human mitochondrial genome database," https://www.mitomaorg/MITOMAP.

[Ph25] "PhyloTree," Maintained by Mannis van Oven; https://www.phylotree.org/tree/index.htm.

[Sc24] Andrew J. Scott, ***The Longevity Imperative: How to Build a Healthier and More Productive Society to Support Our Longer Lives***, Basic Books, (2024); ISBN-13: 978-1541604506.

[Si20] W.A. da Silveira, H. Fazelinia, S. B. Rosenthal, E. C. Laiakis, M. S. Kim, C. Meydan, Y. Kidane, K. S. Rathi, S. M. Smith, B. Stear, Y. Ying, Y. Zhang, J. Foox, S. Zanello, B. Crucian, D. Wang, A. Nugent, H. A. Costa, S. R. Zwart, S. Schrepfer, R. A. L. Elworth, N. Sapoval, T. Treangen, M. MacKay, N, S. Gokhale, S. M. Horner, L. N. Singh, D. C. Wallace, J. S. Willey, J. C. Schisler, R. Meller, J. T. McDonald, K. M. Fisch, G. Hardiman, D. Taylor, C. E. Mason, S. V. Costes, and A. Beheshti, "Comprehensive Multi-omics Analysis Reveals Mitochondrial Stress as a Central Biological Hub for Spaceflight Impact," ***Cell 183***, 1185–1201 (2020); https://doi.org/10.1016/j.cell.2020.11.002.

[Wh25] T. Whipple, "Is mitochondrial therapy the next sports doping scandal?" ***The Times (London)*** February 15 (2025).

Technical Appendices

The intent of this book is to present my assessment of the root cause of human aging and its appropriate treatments, using the simplest language possible. However, there are very relevant technical issues that are too complicated to fit into that format. For this reason, I have included these three Appendices to provide the missing technical information for the interested reader.

J. G. Cramer, *How to Live Much Longer*, Copernicus Books,
https://doi.org/10.1007/978-3-032-17741-4

Appendix A: The Chemical Structures of RNA and DNA

A.1 RNA, Purines and Pyrimidines

What is RNA (ribonucleic acid)? It is an essential building block of all known life. It has the form of a long beaded string of linked sub-molecules, with "beads" made from four possible nucleobases, linked on the string of the ribose-phosphate "backbone". The nucleotide beads of RNA are the two double-ring purines *adenine* (A) and *guanine* (G) and the two single-ring pyrimidines *cytosine* (C) and *uracil* (U). Their chemical structures are shown in Fig. A.1. A closely related pyrimidine, ***thymine*** (T), is used in DNA in place of the uracil in RNA. Thymine has an extra methyl group ($-CH_3$) at the 5th carbon of its pyrimidine ring, while uracil does not. This small change makes thymine more chemically stable than uracil, and this stability is passed on to DNA.

The long RNA chain-like sequence is a variable combination of these nucleotides, arranged in a linear sequence. These nucleobase beads are attached at regular intervals to the string of alternating molecules of the sugar *ribose* ($C_5H_{10}O_5$) and phosphate (PO_4) forming a linear molecular backbone. Here we distinguish between *nucleobases*, which are the basic GCAU molecules, and *nucleotides*, which are nucleobases with ribose and phosphate sub-molecules attached (See Fig. A.1).

The structure of RNA is similar to that of the perhaps more familiar DNA, but there are several key differences: (1) RNA uses the nucleotide *uracil* in

J. G. Cramer, *How to Live Much Longer*, Copernicus Books,
https://doi.org/10.1007/978-3-032-17741-4

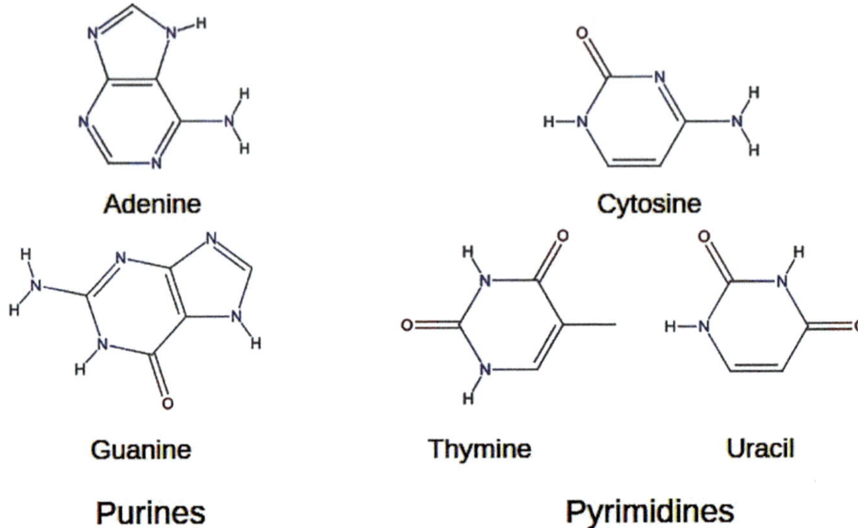

Fig. A.1 Chemical Structure of the GCAU and GCAT Nucleobases. *Credit* JGC, using Wolfram 14.3

the same protein coding role in which DNA uses *thymine*; (2) the RNA backbone uses the sugar *ribose,* while the DNA backbone uses the similar sugar *deoxyribose*, which has almost the same ribose structure but is missing one lower OH group; (3) the RNA chain is usually single-stranded, very flexible, and can readily fold into complex multi-function 3D structures that have many uses, while the DNA chain is fairly stiff and forms a relatively linear double-stranded helix that is more difficult to bend; and (4) RNA is more easily broken down and recycled after short-term use, while DNA is more stable and provides safer long-term storage of genetic information.

A.2 RNA Versus DNA

Figure A.2 shows both the RNA and DNA structures.

In the double helix DNA chain, as Crick and Watson discovered in 1953, each purine always pairs with one particular matching pyrimidine [Wa53]. In particular, the purine guanine always pairs with the pyrimidine cytosine (G + C), and the purine adenine always pairs with the pyrimidine thymine (A + T).

When a single-point mutation occurs that changes the DNA sequence, it is twice as likely that one purine is substituted for the other purine (A↔G) or that one pyrimidine is substituted for the other pyrimidine (C↔T) as it is

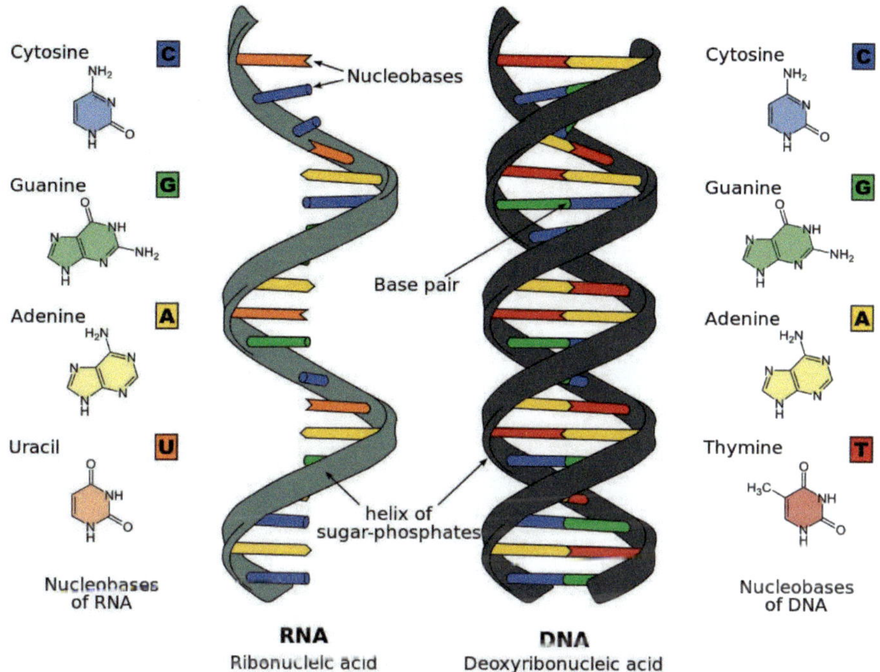

Fig. A.2 The structures of DNA and RNA. *Credit* Sponk CC BY-SA 3.0, via Wikimedia Commons

that a purine is substituted for a pyrimidine or *vice versa*. The former mutation is called a "*transition mutation*" and is primarily caused by transcription errors in replication by the enzyme Pol γ polymerase, while the latter, less common mutation, is called a "*transversion mutation*" and is primarily caused by ROS oxidation. We note that nuclear DNA has a repair mechanism that fixes this type of substitution mutation with 99.9% accuracy, while mtDNA does not have any equivalent mutation repair.

A.3 The RNA Backbone with Nucleotide Beads

The chemical arrangement of the RNA ribose-phosphate chain with one each of the four nucleotides attached is shown in more detail in Fig. A.3. Note that is the nitrogen atom in the purine or pyrimidine ring that bonds to a carbon in the ribose ring.

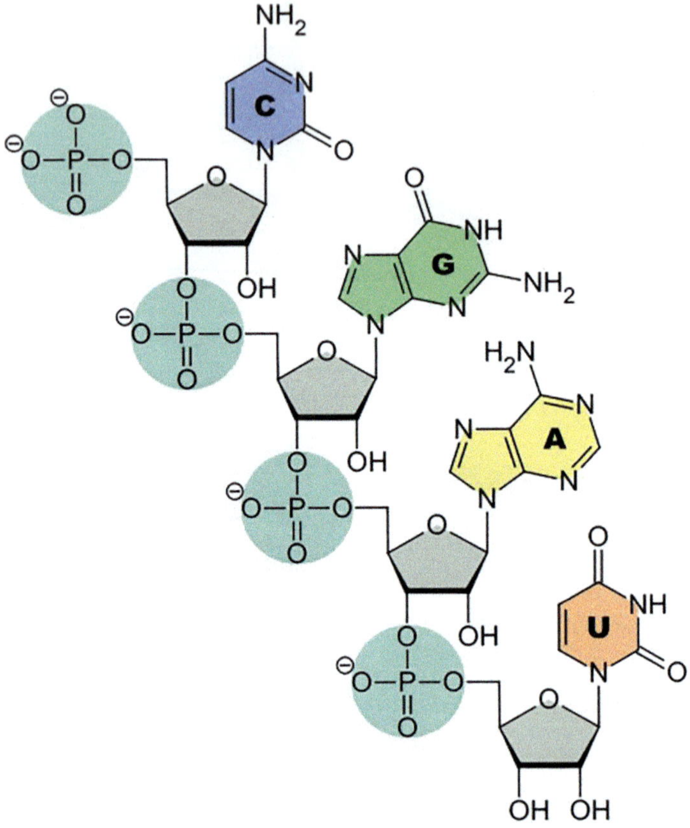

Fig. A.3 The "beaded" RNA chain of ribose (gray pentagons) and phosphate (light blue circles) with attached bases. *Credit* Sponk, Public Domain via Wikimedia Commons

Appendix B: What Actually Damages the mtDNA?

As was discussed in Chap. 3, Sect. 3.1 above, the *mitochondrial free radical theory of aging* developed in the 1950s to 1970s by Harman [Ha56, Ha72], while it had the virtue of bringing into focus age-related damage to mitochondrial DNA, turns out to be a gross oversimplification that has been cast into serious doubt by follow-up experiments and tests. Accumulated mtDNA damage *does* appear to be a primary cause of human aging, but attributing the origin of that damage almost exclusively to direct oxidation by free radicals and, in particular, to damage by the ROS produced in ATP production, is simply wrong and misleading. The ROS oxidation damage is a rather minor direct contributor, often repaired, among the several damage culprits. The causal chain of accumulated mtDNA damage in aging is considerably more complicated.

One of the principal damage culprits is the replication enzyme *mitochondrial DNA Polymerase gamma* (Pol γ). It is the enzyme that is solely responsible for the replication of mtDNA. It is busy with continuous replication activity within the nucleoids of the mitochondrion [Bo08], maintaining a constant supply and outflow of new mtDNA. The Pol γ replication activity takes place at its own pace, sensitive to resource availability but independent of the mitochondrial fission and of the overall cell division cycles. While Pol γ is a relatively high-fidelity polymerase that accurately copies mtDNA with good proofreading, it still makes occasional random errors, possibly triggered by random interactions with other agents in the highly active mitochondrial environment. These intrinsic replication errors are the primary cause of the

J. G. Cramer, *How to Live Much Longer*, Copernicus Books,
https://doi.org/10.1007/978-3-032-17741-4

significant point mutation rate that afflicts mtDNA. Further, when Pol γ encounters a double-strand break (perhaps from ionizing radiation) it also tends to skip a sequence by thousands of base pairs and to make the highly damaging deletion mutation errors.

When making point transcription errors, Pol γ has a preference for making C to T *transition* mutations. On the other hand, the less likely ROS-based oxidative damage tends to make G to T *transversion* mutations, so mtDNA sequencing can reveal the source and rate of the error. The probability of these Pol γ point-mutation replication errors does not seem to depend on cell type or to be age dependent, leading to a linear growth in accumulated mtDNA point mutations. In other words, the point mutations grow at a relatively constant linear rate with age. As we will see, these point mutations, although they relentlessly increase with age, are not the primary source of mtDNA mutation damage and its debilitating consequences for the very old.

Another mtDNA damage culprit is the inadequate mutation repair processes that mtDNA has available. When mtDNA point and deletion mutations do occur, the mitochondrial repair enzymes are often not able to fix them. The mitochondria have inherited their repair processes from their archaic bacterial ancestors, and they are very primitive compared to the excellent processes that actively repair the nuclear DNA (as long as sufficient ATP energy is available to support such energy-expensive repair).

The mtDNA repair machinery that mitochondria do possess is base excision repair (BER). BER is particularly effective at correcting small, non-helix-distorting base lesions such as oxidized bases, deaminated bases (where an amino group ($-NH_2$) has been removed), alkylated bases (where a methyl ($-CH_3$) or ethyl ($-CH_2CH_2$) group has been added), missing bases, and single-strand breaks with damaged ends. But that's all that is available. The more subtle mutations, such as replication errors and base mismatches, insertions, and particularly the deletions in mtDNA, must go unrepaired and therefore will accumulate.

In stark contrast to the linear, even-handedness of point mutation buildup, mtDNA deletion mutations exhibit a distinct cell type preference for accumulating in slowly dividing cells and have an exponential accumulation pattern that changes slope beyond age 65 [Va23]. These difference makes deletion mutations relatively rare in youth but dominant in old age. They are broadly considered to arise from interruption errors during replication, from double-strand breaks, and from the inability of the mtDNA repair system to deal with them.

Double-strand breaks are considered to be a key initiating event for the generation of deletions. These severe lesions can also be induced by various factors, including ROS produced by degenerating mitochondrial ATP production, as well as environmental damaging agents like ionizing radiation or toxic chemicals.

Research also indicates that deletions frequently arise from replication slippage, particularly during the synthesis of the L-strand across the major arc of the mtDNA that starts in the D-loop region (see Chap. 4 above). The unique two-step replication used by mtDNA, where the H-strand is preferentially replicated first, leaves the L-strand exposed as single-stranded DNA for extended periods. This makes it vulnerable to damage and to the formation of abnormal secondary structures and bulges, which can subsequently lead to erroneous cleavage or deletion during replication or repair processes.

The formation of mtDNA deletions often involves the DNA repair machinery itself, when it is engaged in homologous recombination (HR) and non-homologous end-joining (NHEJ). Some deletions exhibit direct repeats at their breakpoints, suggesting generation via HR, while others lack these repeats, implying NHEJ involvement. Faulty proofreading of Pol γ may also play a role when processing the free ends generated by double-strand breaks, preparing them for repair or rejoining. Chemical toxins may form bulky extensions on mtDNA that can cause Pol γ to stall during replication. This can interfere with mtDNA replication and can lead to deletions or lower mtDNA copy number.

Detailed nanopore sequencing of mtDNA as a function of subject age has revealed an exponential growth in deletion mutations in the elderly, as discussed below [Va23]. One plausible hypothesis for explaining this exponential accumulation of mtDNA deletions with age is the "replicative advantage" that mtDNA with deletion mutations have due to the smaller size of the DNA ring to be replicated. It is well known in the laboratory PCR replication of DNA that the shorter DNA segments are replicated exponentially-with-inverse-length faster than the long DNA segments, presumably because short segments require fewer interactions with the replicating polymerase enzymes and require less of the time-consuming transport and attachment of bases to the DNA backbone before completion. The same asymmetry seems to also be present in the Pol γ replication of mtDNA within the mitochondrion. There is sequencing evidence that shorter mtDNA rings having large deletions (> 3,000 bp) are replicated faster and accumulate faster than longer intact mtDNA rings having few or no deletions.

This implies that, in the ongoing and continuous replication of mtDNA within the mitochondrion, the more deletion-damaged mtDNA rings will have a distinct replicative advantage. Over time, this can result in the buildup of very high fractions of deletion-damaged mtDNA in cells or tissues of those

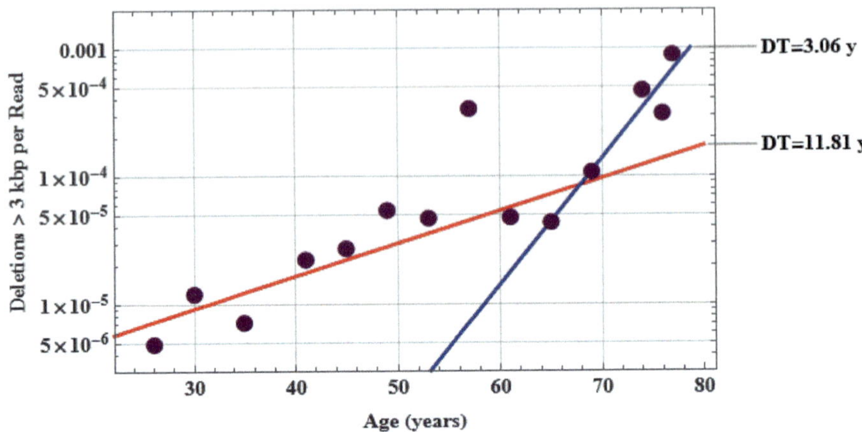

Fig. B.1 Deletion Mutations > 3 kbp versus Age. *Credit* data from [Va23] used with permission, plot and fits by JGC using Wolfram 14.3

of advanced age, particularly if mitophagy has fallen off in efficiency and does not detect and remove the "quiet" low-membrane-potential mitochondria that are dominated by mtDNA with such deletions. It has been clinically observed that mtDNA deletion damage can reach over 50% in some neurons of the aged mid-brain *substantia nigra,* and even higher levels in patients with Parkinson's disease.

In 2023, a group based in Cologne, Germany, and UCLA [Va23] performed precision nanopore sequencing analysis on the mtDNA taken from the *vastus lateralis* muscle of 15 males ranging from 20 to 81 years. Data from their results for mtDNA deletions with more than 3,000 base pairs missing, as shown in Fig. B.1, indicate a definite trend for exponential growth. There is more random variation in shorter deletions, which in any case replicate more slowly. They found that deletion damage in the mtDNA genome begins to increase *exponentially* beyond about age 30.

However, there also seems to be a slope increase in the data at about age 65. Here, their data (obtained with permission) has been re-plotted on a log scale and separately fitted with two exponentials: up to age 65 (red line) and age 65 and older (blue line). (The wild point at 57 years was not included.) The fits indicate that the doubling time (DT) before age 65 is 11.81 years and the doubling time after age 65 is 3.06 years. In other words, for men over 65, the quantity of large deletion mutations (at least in their muscle cells and probably in many other organs) increases by a factor of about 9.7 every 10 years, with severe debilitating consequences.

Why do such deletion mutations have exponential growth? Because, as discussed above, the deleted mtDNA rings that have over 18% of their length missing are significantly shorter, so that the replication process can go at least 21% faster than replication of a full-length mtDNA loop. This allows mtDNA with such deletions to proliferate more rapidly than the full or point-mutated mtDNA within the cell. That creates a powerful positive feedback loop, with the fraction of deletion mutations growing with each replication cycle, leading to the observed rapid age-dependent exponential accumulation of large deletion mutations for humans beyond age 65, mutations that double in number about every 3 years.

For this reason, it appears that mtDNA deletion mutations are the primary driver of the age-related symptoms and diseases of human old age. In other words, it is not the linearly increasing *point* mutations, but rather the exponentially growing and equally damaging *deletion* mutations of the mtDNA that drive the debilitation, decline, and onset of age-related diseases of human old age. Even with a small beginning, an exponential increase can always outrun a linear increase, given enough time.

Appendix C: The Bottleneck: How the mtDNA Winnowing Works

There is a winnowing process during human reproduction that culls the vast majority of the mother's damaged mtDNA and mitochondria and effectively resets the mtDNA in the developing embryo to a nearly perfect and undamaged state. The process begins with the mother's primordial germ cells (PGCs). These are the precursors to oocytes (unfertilized eggs), which initially contain only 10–100 mtDNA copies. As the PGCs proliferate into oogonia and then into primary oocytes during development, the mitochondrial pool is partitioned among the dividing cells. The "bottleneck" arises from the way in which these mitochondria are segregated and replicated. During these early and later stages of oocyte creation, there's a random segregation of mitochondria, with clearance of the unlucky oocytes that end up containing too much damaged mtDNA.

A significant number of the oocytes undergo programmed cell death (atresia) during development, providing a mechanism for "cellular-level selection", eliminating oocytes that have accumulated damaged mitochondria. A crucial part of the selection process occurs within an internal structure called the *Balbiani body* of the developing oocyte. The Balbiani body is where organelles, including good mitochondria, are concentrated. To ensure the selection of high-quality mitochondria for the oocyte, the high-quality mitochondria are selectively routed into the Balbiani body and pooled there, while damaged ones are excluded and collected in mitochondria targeted for mitophagy. In the early growing oocytes, the mtDNA begins replicating, and the mitochondrial count per cell goes up to 1,000 to 10,000. In the mature

J. G. Cramer, *How to Live Much Longer*, Copernicus Books,
https://doi.org/10.1007/978-3-032-17741-4

ovulated oocyte, the mitochondrial count per cell goes up to 100,000, which is the high point, with each of the mitochondria busily churning out ATP to prepare for the large energy needs of embryo creation.

We note the the replicative advantage of mtDNA having large deletion mutations, discussed in Appendix 2 above, also applies to the rapid mtDNA replication in the oocyte. If a mitochondrion containing deletion-mutated mtDNA happens to slip by the selection process described above, it can come to dominate the mtDNA copies in the mature ovulated oocyte, with disastrous consequences for the embryo in the form of miscarriage, stillbirth, or mitochondrial genetic diseases.

After fertilization, in the initial cell divisions of the disconnected embryo, the zygote's mitochondria count is initially about 100,000. At each subsequent division, the blastomeres' mitochondrial count drops by a factor of 2, as the mitochondria are shared between dividing cells without much new replication.

However, when the cell count reaches around 32, each cell having around 3,000 mitochondria, mitochondrial fission resumes, and the new cells downstream each have around 3,000 mitochondria. This ensures that all the cells of the developing embryo, and ultimately the adult organism, are populated with a sufficient number of healthy mitochondria to supply the needed ATP energy. The total number of mitochondria stays around 3,000–4,000 per mature embryonic cell, which then replicate to provide for the billions of cells in the emerging human. However, all of the resulting mtDNA were derived from 10 or so selected copies of the mother's mitochondrial genome.

References

[Bo08] D. F. Bogenhagen, D. Rousseau, and S. Burke, "The Layered Structure of Human Mitochondrial DNA Nucleoids," ***Journal Of Biological Chemistry 283*** (6), 3665–3675 (2008).

[Ha56] D. Harman,. "Aging: A theory based on free radical and radiation chemistry," ***Journal of Gerontology 11*** (3), 298–300 (1956).

[Ha72] D. Harman, "The biologic clock: the mitochondria?", ***Journal of the American Geriatrics Society 20*** (1972).

[Va23] A. R. Vandiver, A. N. Hoang, A. Herbst, C. C. Lee, J. M. Aiken, D. McKenzie, M. A. Teitell, W. Timp, and J. Wanagat, "Nanopore sequencing identifies a higher frequency and expanded spectrum of mitochondrial DNA deletion mutations in human aging," ***Aging Cell 22*** (6) (2023); https://doi.org/10.1111/acel.13842

[Wa53] J. D. Watson and F. H. C. Crick, "Molecular Structure of Nucleic Acids: A Structure for Deoxyribose Nucleic Acid," ***Nature 171***, 737–738 (1953).

J. G. Cramer, *How to Live Much Longer*, Copernicus Books,
https://doi.org/10.1007/978-3-032-17741-4

Index

J. G. Cramer, *How to Live Much Longer*, Copernicus Books,
https://doi.org/10.1007/978-3-032-17741-4